Amit Setia, Denis O. Krause

E.coli: Probiotic for Piglet Diarrhoea

Amit Setia
Denis O. Krause

E.coli: Probiotic for Piglet Diarrhoea

Escherichia Coli K88+ post Weaning Diarrhoea

VDM Verlag Dr. Müller

Imprint

Bibliographic information by the German National Library: The German National Library lists this publication at the German National Bibliography; detailed bibliographic information is available on the Internet at http://dnb.d-nb.de.

Cover image: www.purestockx.com

Publisher:
VDM Verlag Dr. Müller Aktiengesellschaft & Co. KG, Dudweiler Landstr. 125 a, 66123 Saarbrücken, Germany,
Phone +49 681 9100-698, Fax +49 681 9100-988,
Email: info@vdm-verlag.de

Produced in USA and UK by:
Lightning Source Inc., La Vergne, Tennessee, USA
Lightning Source UK Ltd., Milton Keynes, UK
BookSurge LLC, 5341 Dorchester Road, Suite 16, North Charleston, SC 29418, USA

ISBN: 978-3-639-02627-6

Abstract

Setia, Amit. M.Sc., The University of Manitoba, April, 2007. Selection of _Escherichia coli_ K88$^+$ specific probiotic strains of _E. coli_ from environmental isolates for post-weaning piglets. Advisor: Denis O Krause.

Enterotoxigenic _Escherichia coli_ (ETEC) cause diarrhea in domestic animals and humans. ETEC diarrhea poses a major threat to the swine industry in North America as well as many other continents of the world. Significant financial losses are suffered as a result of piglet mortality, the cost of poor weight gain in surviving piglets as well as medication. Thus research in an economically important disease like post-weaning diarrhea (PWD) due to _E. coli_ K88$^+$ is valuable to the swine industry. In order to cause diarrhea, ETEC must adhere, proliferate, and produce toxins in the host intestines. ETEC attach to intestinal mucous membranes via adhesins, such as K88 (F4), K99 (F5) and 987P (F6) fimbriae. ETEC strains bearing the K88 fimbrial antigens or _E. coli_ K88$^+$, are the most common causative agents of colibacillary diarrhea in neonatal and early weaned piglets and is an exclusive swine pathogen. Clinical signs of diarrhea from ETEC infection are induced by secretion of enterotoxins after colonization by ETEC.

Sub-therapeutic antibiotics are routinely added to feed or drinking water to prevent ETEC diarrhea in North American swine farms. Antibiotics have been demonstrated to improve the growth rates and feed conversion ratios up to 16%, and 6%, respectively, in nursery piglets. In Canada, antibiotics commonly used are chlortetracycline-sulphamethazine-penicillin and tylosin-furazolidone which are added to creep and starter rations. The use

i

of antibiotics in farm animal production systems has been blamed for the emergence of multi-drug resistant pathogens in human medicine and has limited their use in farm animals. The increased risk of antimicrobial resistance has compelled producers to find alternatives to antibiotics. Many different approaches are available, including the use of oral vaccination with purified fimbriae, organic acids, egg yolk antibodies, zinc supplementation; spray dried plasma, phage therapy, breeding F4 and F18 resistant animals, and finally, probiotics and prebiotics.

Direct fed microbes, or probiotics, are a potential alternative to antibiotic feed supplements for controlling multiplication and attachment of pathogens in the intestines of pigs. Most studies have successfully used Gram-positive lactic acid bacterial strains in probiotic feed supplements for pigs. These lactic acid bacteria are not known to possess a direct bactericidal effect on diarrhea causing *E. coli* K88$^+$. A probiotic with antimicrobial properties against target pathogens and without being harmful to its host animal can be a better alternative to antibiotics.   This approach is also more cost effective and environmentally friendly, as it circumvents the problem of emerging antibiotic-resistant organisms. In this project, an *in vitro* study was carried out to select *E. coli* K88$^+$ inhibitory probiotic strains from environmental *E. coli* isolates based on the property of *E. coli* known as colicinogeny along with the absence of toxin production, and the ability to utilize specific carbohydrate sources.

The objectives of this thesis were: (i) to select specific colicinogenic *E. coli* inhibitory to *E. coli K88*$^+$ from environmental isolates; (ii) to identify colicins that were effective

ii

against *E. coli* bearing K88 type fimbriae; (iii) to characterize the identified isolates (iv) to identify prebiotics, or carbohydrates specifically utilized by potentially probiotic *E. coli* and not by the pathogen and (v) to evaluate efficacy of selected isolates against *E. coli* K88$^+$ *in vitro*.

In this study, we screened 463 strains of *E. coli* isolated from hog manure-amended soil (37), pig feces (35), cattle rumen fluid (33) and cattle feces (358) for production of colicins against an indicator strain of *E. coli* K88$^+$ (2-12) obtained from Dr. Carlton Gyles (University of Guelph). Colicins effective against *E. coli* K88$^+$ were identified by testing 18 reference *E. coli* K-12 constructs with predefined colicin plasmids against clinical isolates of *E. coli* K88$^+$. Toxin genes for heat stable (STa and STb), heat labile (LT) and Vero (VT1 and VT2) toxins were detected by using PCR reactions.

From the plate inhibition test we concluded that colicin E3, E4, E5, E9, N, K, and Ia inhibited all 12 clinical strains of *E. coli* K88$^+$. Approximately 26% of strains from cattle and pig feces produced colicins against *E. coli* K88$^+$, but more than 43% of strains from soil could inhibit *E. coli* K88$^+$. Of the 121 colicinogenic strains isolated, 32 were positive for VT1, 7 for VT2, 23 for STa, 3 for STb and none for LT. The most colicinogenic of 56 toxin-free isolates were selected for further screening. We selected 14 of the most colicinogenic isolates for utilization of carbon sources from a panel of 49 different carbohydrates. . There were differences in utilization of 15 different carbohydrates among the selected environmental *E. coli* and *E. coli* K88$^+$. Two of the selected environmental isolates could utilize starch and inulin as the sole carbohydrate source, whereas none of five randomly tested *E. coli* K88$^+$ strains could utilize these carbohydrates. In the present study it was demonstrated that *in vitro* competition between *E. coli* K88$^+$ and colicinogenic isolates resulted in significant inhibition of *E. coli* K88$^+$ in the presence either of potato starch or inulin as well as in a standard growth medium. From this study it was concluded that there was strong *in vitro* evidence that the use of a *E. coli* K88$^+$ specific colicinogenic probiotic, formulated with starch as a specific prebiotic, would control *E. coli* K88$^+$-induced diarrhea in weaned piglets.

Acknowledgments

I would like to thank my supervisors, Dr. D O Krause and Dr. J D House for their excellent support, advice and guidance throughout my master's program and thesis writing. Many thanks are extended to Dr. C M Nyachoti and Dr. R A Holley for their guidance and advice on all the work time to time. Technical assistance, encouragement and friendship of Dr. R Kotlowski and S K Bhandari is greatly appreciated. I greatly acknowledge the department of Animal Science for providing their facilities. Thanks also to Cathy Plouffe, Ilse Frank, Margaret Ann and Claire Hutchinson for helping with graduate student matters during the entire program.

I would like to thank all my friends, colleagues and technicians in Animal Science for their company and all sorts of help during these years. Special thanks to my friends, Bai Xu, Tyra, Fuqin, Wei, Shun, Gian-carlo, Florence, Juan, Elijah, Ehsan and Shadi. Many thanks to my friends who have been more like family, Amarbeer, Obi, Natalie, Ainsley and Stephanie for their excellent company, encouragement and support. Finally, my sincere thanks and appreciation goes out to 'all' members of my family, especially mum, papa, Jeetu, Kunnu and Ruby for their endless support and for my wife, Shalini for her support and patience.

Dedication

This work is dedicated to my late grandparents who stimulated my interest in animal science and inculcated in me the importance of education and hard work.

Foreword

A part of this thesis has been written in manuscript format. The thesis is organized with an abstract of the thesis, general introduction and a review of the literature prior to the manuscript, followed by a general discussion and conclusions. The format used for writing this thesis is that of the journal *"Microbiology"* published by the Society of General Microbiology (United Kingdom).

The authors and title of this manuscript are:

Amit Setia, Sanjiv K. Bhandari, James D. House, Charles M. Nyachoti, & Denis O. Krause (2007). Development of *Escherichia coli* K88[+] specific probiotic strains for post-weaning piglets. (in preparation)

Table of Contents

Abstract... i
Acknowledgements.. v
Dedication.. vi
Foreword.. vii
Table of Contents... viii
List of Tables.. ix
List of Figures.. x
Table of Abbreviations.. xi
1.0 General Introduction 1
2.0 Literature Review... 3
    2.1. Diarrheagenic *Escherichia coli* in Pigs.......................... 3
    2.2. Alternatives to Antibiotics against *E. coli* K88[+] 6
        2.2.1 Egg yolk antibodies... 7
        2.2.2 Spray dried animal plasma (SDAP)....................... 8
        2.2.3 Organic acids.. 9
        2.2.4 Vaccination... 10
        2.2.5 Transgenic plants.. 11
        2.2.6 Probiotics... 12
            2.2.6.1 *E. coli* as a probiotic............................... 15
        2.2.7 Prebiotics... 16
    2.3 *E. coli* Bacteriocins- Colicins....................................... 18
        2.3.1 Colicin gene 20
        2.3.2 Classification of colicins..................................... 21
        2.3.3 General structure of colicins................................ 22
        2.3.4 Colicins and pathogenic *E. coli*.......................... 23
        2.3.5 Colicins summary... 25
3.0 Hypotheses and Objectives.. 26
4.0 Manuscript: Development of a colicinogenic *Escherichia coli* K88[+] specific probiotic strain for post-weaning piglets.......................... 27
    4.1 Abstract.. 28
    4.2 Introduction... 29
    4.3 Methods.. 31
    4.4 Results.. 36
    4.5 Discussion.. 39
    4.6 Acknowledgements... 44
    4.8 Tables and Figure... 45
5.0 General Summary 56
6.0 Conclusion and Future research... 59
7.0 Literature Cited.. 61
8.0 Appendices.. 78
    8.1 Appendix-A.. 78
    8.2 Appendix-B.. 80
    8.3 Appendix-C.. 87

List of Tables

No.	Tables in manuscript	Page
4.1	List of reference strains used in current study and their characteristics.	45
4.2	PCR primers designed and used in this study and their sequences.	46
4.3	Sources of environmental *E. coli* isolates and distribution of colicinogeny Among isolates.	47
4.4	Phenotypic characterization of selected most colicinogenic environmental *E. coli*.	48
4.5	Phenotypic characterization of *E. coli* K88$^+$ strains.	50

No.	Tables in Appendices	Page
B-4.1	List of substrates in API 50 CH identification kit	86

List of Figures

No.	Figures in manuscript	Page
4.1	*In-vitro* competition assay of UM-2 and *E. coli* K88$^+$ 2-12.	52
4.2	*In-vitro* competition assay of UM-7 and *E. coli* K88$^+$ 2-12.	53
4.3	Averages of zones of inhibitions of top 14 isolates from selected potential probiotic *E. coli*.	54
4.5	Cluster analysis of enterotoxins from colicinogenic environmental strains	55

No.	Figures in appendices	Page
A-2.1	Structure of colicin gene cluster on plasmid.	78
A-2.2	Phylogenetic relationship among various pore forming and nuclease colicin groups based on C-terminal 300 amino acids of colicin proteins.	78
A-2.3	The domain structure of colicins.	79
A-2.4	Domain structures and their lengths in colicins E1, Ia and A.	79
B-4.1	Diagrammatic representation of modified procedure used for detection of colicinogeny in environmental isolates of *E. coli*.	80
B-4.2	Flow chart for experimental approach used in selecting potentially probiotic environmental *E. coli* isolates.	81
B-4.3	Flowchart for characterization of 14 selected environmental *E. coli* isolates.	82
B-4.4	Plate test for detection of colicin production against *E. coli* K88$^+$.	83
B-4.5	Representative pictures of agarose gel electrophoretograms.	84
B-4.6	Standard results for API 50 CH. A. Negative reactions B. Positive reactions on API 50 CH strips. (Biomerieux)	85
C-4.1	Primer sequence for Fimbriae K88- 439 bp	87
C-4.2	Primer sequence for Colicin E1- 397 bp	88
C-4.3	Primer sequences for Colicin A, N, S4- 225 bp	89
C-4.4	Primer sequences for Colicin M- 556 bp	90
C-4.5	Primer sequences for Colicin V- 400 bp	91
C-4.6	Primer sequences for Colicin U and colicin Y- 243 bp	92
C-4.7	Primer sequences for Colicin Ia, Ib- 385 bp	93
C-4.8	Primer sequences for Colicin B, D, D157- 138 bp	94
C-4.9	Primer sequences for Colicin Group E (E2, E3, E3a, E4, E5, E6, E7, E8, E9) – 219 bp	95
C-4.10	Primer sequences for Colicin 5, 10, K- 803 bp	96

Table of Abbreviations

Abbreviation	Definition
bp	Base pairs
CFU	Colony forming units
dNTP	Deoxyribonucleoside triphosphates
ETEC	Enterotoxigenic *Escherichia coli*
FOS	Fructooligosaccharides
GIT	Gastrointestinal tract
LB agar	Luria Bertani agar
LB broth	Luria Bertani broth
LBm	Luria Bertani agar with mitomycin-C
LT	Heat labile enterotoxin
PBS	Phosphate buffered saline
PWD	Post-weaning diarrhea
STa	Heat stable enterotoxin a
STb	Heat stable enterotoxin b
Taq	*Thermus aquaticus* DNA polymerase enzyme
Tm	Melting temperature (as applied to primers)
VT-1	Vero toxin-1
VT-2	Vero toxin-2

1.0 GENERAL INTRODUCTION

Post-weaning diarrhea and the cost associated with loss of production often present a major problem to farms wishing to practice antibiotic free swine production (Amezcua *et al.*, 2002; van Beers-Schreurs *et al.*, 1992). Increasing antimicrobial resistance in pathogens has lead to the banning of antibiotics as sub-therapeutic feed supplements in swine production to control post weaning diarrhea in many countries (Amezcua *et al.*, 2002; Nagy & Fekete, 2005). In Canada, commonly used antibiotics such as chlortetracycline-sulphamethazine-penicillin and tylosin-furazolidone are added to creep and starter rations to control the problem of post weaning diarrhea, however due worldwide concern over the use of antibiotics in animal agriculture and their possible contribution to antibiotic resistance, the development of effective alternatives to conventional antibiotics is immediately needed to protect swine from *E. coli* K88$^+$ infections.

The use of probiotics has been shown to be effective in reducing diarrhea caused by ETEC K88 strains (Blomberg *et al.*, 1993a; Jin *et al.*, 2000; Revajova *et al.*, 2000). However, the consistency of results has always been questionable (Nagy & Fekete, 2005). Traditionally, Gram-positive lactic acid bacteria have been used as probiotic strains (Holzapfel *et al.*, 2001). Although use of Gram-negative species as a probiotic is rare in swine production, there have been recent studies on the use of fungi, yeasts and some Gram-negative microorganisms as probiotics (Isolauri *et al.*, 2002; Schroeder *et al.*, 2006). Other studies have confirmed the use of *E. coli* as probiotics in human conditions (Kruis *et al.*, 1997; Malchow, 1997; Lodinova-Zadnikova & Sonnenborn, 1997). One of

1

the best examples is *E. coli* strain Nissle 1917 which is commonly used as a probiotic to treat Crohn's disease and ulcerative colitis in humans (Kruis *et al.*, 1997; Malchow, 1997).

Bacteriocins are biologically active protein molecules with a bactericidal mode of action and have a narrow activity spectrum (Lakey *et al.*, 1994). *E. coli* colicins are the most extensively studied type of bacteriocins. To date, no studies have utilized colicinogenic *E. coli* as a competitive exclusion system for swine, even though colicins of *E. coli* have been shown to protect animals from gastrointestinal diseases (Schamberger & Diez-Gonzalez, 2002). Colicin producing *E. coli* are logical tools to eliminate ETEC *E. coli* K88$^+$ from gastrointestinal tract of weaned piglets. Little research has been done to investigate the effect of colicins in farm animal gastrointestinal diseases particularly infections in pigs caused by *E. coli*. It is therefore worth looking into finding *E. coli* K88$^+$ specific colicinogenic probiotic strains as an alternative to antibiotic additives for weaned piglets.

2.0 Literature Review

2.1 Diarrheagenic *Escherichia coli* in Pigs

There are six major groups of diarrheagenic *Escherichia coli* (*E. coli*) : enterotoxigenic *E. coli* (ETEC), enterohemorrhagic *E. coli* (EHEC), enteroinvasive *E. coli* (EIEC), enteropathogenic *E. coli* (EPEC), enteroaggragative *E. coli* (EAggEC) and diffusely adherent *E. coli* (DAEC). These groups have been extensively investigated including their mechanism of pathogenesis and the development of diagnostic techniques (Nataro & Kaper 1998). The focus of this review will be on the swine pathogenic ETEC strain, *E. coli* K88$^+$.

Post-weaning diarrhea (PWD), also called post-weaning enteric colibacillosis, is characterized by diarrhea, dehydration, weight loss, and mortality in newly weaned piglets (Kahn, 2005). Disease is usually caused by *E. coli* that carry F4 (K88) or F18 adhesins, also known as fimbriae, or sometimes referred to as pili (Kahn, 2005; Fairbrother *et al.*, 2005). ETEC strains bearing the K88 fimbrial antigens are the most common causative agents of colibacillary diarrhea in neonatal and early weaned piglets and are an exclusive swine pathogen (Yokoyama *et al.*, 1992; Blomberg *et al.*, 1993b Cassels & Wolf, 1995; Marquardt *et al.*, 1999). Diarrhea due to *E. coli* K88$^+$ is a major economically important problem for the swine industry in North America and world wide (Amezcua *et al.*, 2002; van Beers-Schreurs *et al.*, 1992). Newly weaned piglets are especially predisposed to ETEC diarrhoea due to multiple factors. These piglets have low levels of digestive enzymes, poor innate immunity, declining passive immunity from

3

colostrum, and stress due to changed environment (Palmer & Hulland, 1965; Nagy & Fekete, 1999).

ETEC attach to intestinal mucous membranes by the application of adhesins like K88 (F4), K99 (F5), and 987P (F6) fimbriae (Nagy *et al.*, 1990). Fimbriae are thin flexible hair like structures on the surface of *E. coli* with a diameter of 3-7 nm, and differ from pili which are rigid hollow structures with a diameter of 7-8 nm (Nagy & Fekete, 1999; de Graaf & Mooi, 1986). ETEC fimbriae make pathogens host specific as they play an important role in the attachment of *E. coli* to their respective host receptors on the intestinal mucosa (Verdonck *et al.*, 2004). K88 expressing *E. coli* triggers disease exclusively in newly weaned pigs (Cassels & Wolf, 1995) causing diarrhea, delaying growth, and reducing production on swine farms (Bosi *et al.*, 2004). There are three antigenic variants of the K88 fimbrial adhesins, K88ab, K88ac, and K88ad. The genes coding both fimbriae and toxins are located on plasmids (deGraaf *et al.*, 1984).

Adhesion by *E. coli* K88 to intestinal mucosa is likely a prerequisite for colonizing the gastrointestinal tract (GIT) (Gaastra & deGraaf, 1982). However, not all young pigs are affected by *E. coli* producing these types of fimbriae. Expression of intestinal brush border mucin type glycoprotein (IMTGP) was highly correlated with susceptibility to K88ab and K88ac suggesting the development of possible resistant breeds of pigs (Francis *et al.*, 1998).

Generic ETEC which includes a number of serotypes other than K88 cause colibacillary diarrhea in domestic animals and humans (Padhye & Doyle, 1991; Nagy & Fekete,

2005). It is an acute diarrheal disease which occurs in newborn calves, lambs, and piglets. ETEC also causes colibacillary toxemia or the increase of fluid secretion during diarrhea through the release of endotoxins. In pigs, colibacillary toxemia may be exhibited as weaner shock syndrome, hemorrhagic enteritis, and edema disease. Absorption of endotoxin from the GIT is proposed to play a role in the pathogenesis of shock in weaner syndrome.

E. coli enterotoxins are large (heat-labile toxins) or small (heat stable toxins) molecular weight proteins that cause hyper secretion of water and electrolytes from enterocytes leading to watery diarrhea. Some ETEC may produce more than one toxin (Nagy & Fekete, 1999) that are mostly combinations of heat-labile toxin (LT) and heat stable toxin-b (STb) or a combination of LT, heat stable toxin-a (STa) and STb (Celemin *et al.*, 1995; Amezcua *et al.*, 2002). The mechanism of action of LT and ST enterotoxins is that they cause increased secretion of water, Na^+, Cl^- and decrease fluid absorption that in turn causes dehydration, acidosis and diarrhea (Nagy & Fekete, 2005). Ameczua *et al.* (2002) reported outbreaks of post weaning *E. coli* diarrhea due to toxin producing *E. coli* $K88^+$ strains in piglets in southern Ontario. They also reported multi- drug resistance in these pathogens. Noamani *et al.* (2003) found the STa gene in 92% of isolates from post-weaning diarrhea cases in Ontario. Another of the major toxins produced in this disease is vertoxin 2e (VT) and is produced by the O antigens on the lipopolysaccharide of *E. coli*. These O antigens are highly heterogeneous and consist of a number of serotypes including O8, O45, O138, O139, O141, and O149 (Nagy & Fekete, 2005; Schultz & Strokbine, 2005).

It is a curious observation that *E. coli* are normally present in the digestive tract of pigs and other animals, and even pathogenic serotypes may be present without causing disease. This is because the presence of a pathogen is one of several (but not an exclusive) prerequisites for pathogenesis. For instance, the brush border membranes of pig intestines have six receptor proteins with molecular weights ranging from 27 to 94 kDa that bind *E. coli* K88ab fimbriae (Caloca *et al.,* 2002), but these receptors are only present at between 3 and 8 weeks of age and thereafter disappear from the digestive tract. Additionally, pigs that are of the required age and are expressing receptors do not necessarily become ill. Compromise of the innate immune systems by events like cold or social stress is an important trigger of pathogenesis. Good animal husbandry can usually prevent occurrence of significant disease in these young pigs.

2.2 Alternatives to antibiotics against *E. coli* K88[+]

In recent years, several alternatives to antibiotics have been tested to control post weaning diarrhea in piglets. These alternatives include egg yolk antibodies, oral vaccination with purified antigens, spray dried animal plasma, organic acids, and transgenic plants, however, no single method has been completely satisfactory in terms of disease prevention and cost efficacy. Owusu-Asiedu *et al.* (2003) studied the effect of spray dried porcine plasma, pea protein isolate with egg yolk antibody, zinc oxide, fumaric acid added as well as antibiotics in piglets in relation to *E. coli* K88[+] diarrhea and

6

found these alternatives to antibiotics to be effective. The main categories of currently accepted alternatives are included in this part of the chapter.

2.2.1 Egg yolk antibodies

Egg-yolk antibodies produced from fimbrial antigen-immunized laying hens were found to inhibit the binding of *E. coli* K88$^+$ to mucosal receptors. Egg-yolk antibodies against K88 fimbria when added to isolated pig mucin could prevent the adhesion of *E. coli* K88$^+$ (Jin *et al.*, 1998). Almost 100% of 21 day old piglets challenged with *E. coli* K88$^+$ survived with only a low incidence of diarrhea when anti-K88 antibodies prepared from egg yolk was used as a therapeutic agent (Marquardt *et al.*, 1999). In fact there are many studies that have specifically reported an effect with egg-yolk antibodies to K88. More recently, contradictory results to the above study were reported by Chernysheva *et al* (2003) in a field trial with two swine herds that had a history of K88 diarrhea. These authors evaluated the efficacy of a commercial product containing egg yolk antibodies specific for control of *E. coli* K88$^+$ incorporated into nursery feed and found no effect of the treatment on prevalence of diarrhea or mortality.

In another study, Harmsen *et al.* (2005) tested oral passive immunotherapy using F4 (K88) fimbriae specific llama single domain antibody. These antibodies successfully inhibited attachment of ETEC F4 to intestinal brush border; however they poorly protected experimentally infected piglets (Harmsen *et al.*, 2005). This raises the possibility that the antibody to a very specific strain of K88 was produced for the product and that to be effective in the field, where the K88 genotype is likely to be heterogenous, a higher degree of polyclonality is required in the egg yolk supplement.

2.2.2 Spray dried animal plasma (SDAP)

SDAP contains proteins including IgG which can provide oral passive immunity to weaned piglets. In various studies, inclusion of SDAP in the diet of early weaned pigs reduced the incidence and severity of enteric infections (Owusu-Asiedu et al., 2003; Bosi et al., 2004). Use of an animal byproduct has its own limitation due to pathogen transmittance especially that of viruses and scrapie as learnt from bovine spongiform encephalopathy (BSE) outbreaks in Europe (Bradley et al., 2006).

Theoretically, the use of an E. coli K88[+] challenge model has many advantages when investigating supplements such as SDAP in diets as it removes the possibility that animals are not infected by E. coli K88[+] and also chances of non-specific response can be excluded (Owusu-Asiedu et al., 2003; Bosi et al., 2004). However, these challenge models also have their shortcomings. Many researchers (Bosi et al., 2004; Yi et al., 2005) have noted that protein source has a major impact on animal performance in the E. coli K88[+] challenge model. Bosi et al. (2004) compared fish protein to spray dried animal plasma (SDAP) in a challenge model with E. coli in medicated and non-medicated feed. They observed a significant animal performance response for protein source in which SDAP was the superior ingredient (Bosi et al., 2004). However, no improvement in growth performance was obtained when antibiotics were added to the SDAP-based diet

(Bosi *et al.*, 2004). This result was also true for the fish protein-based diets, but medication tended to have a beneficial effect (Bosi *et al.*, 2004). Yi *et al.* (2005), compared SDAP, glutamine, and a diet without either in an *E. coli* challenge model. They found no effects in growth performance (Yi *et al.*, 2005). The performance response they observed was in their positive control that was not challenged with *E. coli* K88^{+} (Yi *et al.*, 2005).

2.2.3 Organic acids

Organic acid salts like fumarate, propionate, and citrate are commonly included in nursery pig diets as supplements as a means of controlling post-weaning diarrhea (Verstegen & Williams, 2002). The proposed mechanism is based on the fact that at the age pigs are typically weaned, production of HCl in the gastric juice in stomach is insufficient to reduce pH of the gut lumen to values that are typically inhibitory to *E. coli* (Verstegen & Williams, 2002). These pH's are generally considered to be in the range of 2 to 4, and this range is not normally reached in the stomach of pigs until the age of 7 to 8 weeks of age. The addition of organic acids helps to overcome the buffering capacity of feed and help reduce the pH in the stomach.

Cherrington *et al.* (1991) demonstrated that propionic acid at pH 5.0 in concentrations of 0.5-0.7 mol. L^{-1} kills 90% of *E. coli* K-12 and *Salmonella* spp. within 60 min when compared to over 3 hours if formic acid was used at the same concentration. Gluconic acid (Biagi *et al.*, 2006) can improve growth performance in weaned piglets and

9

positively influences the composition and activity of microflora. Owusu-Asiedu *et al.* (2003) demonstrated a reduction in severity of K88-induced diarrhea by adding fumaric acid to a pea protein isolate-based diet for weaned piglets. Benzoic acid is also known to exert a strong antimicrobial effect on Gram-negative bacteria in the piglet intestinal tract. It also enhances nitrogen retention and weight gain in benzoic acid fed pigs (Kluge *et al.,* 2006).

In view of the fact that one of the major virulence factors of *E. coli* is acid resistance, it is difficult to conclude that reduced pH is the only role that organic acids have in the gut. The pH reducing effects of organic acids probably also result in the prolifieration of other lactic acid bacteria (Pederson *et al.,* 2004). A number of studies have shown that *Lactobacillus* spp. increase in the stomach of pigs consuming organic acids (Verstegen & Williams, 2002). In this sense, organic acids can also be seen as having a prebiotic effect (see discussion below) and organic acids themselves can serve as electron sinks in anaerobic metabolism.

2.2.4 Vaccination

The success of vaccination against PWD depends on the production of specific antibodies against antigens. Van den Broeck *et al.* (1999) applied an oral vaccine using purified F4 fimbriae to immunize pigs. Bertschinger *et al.* (2000) demonstrated the efficacy of a live oral vaccine containing F18 fimbriae to protect weaned pigs against PWD and edema disease. In general, however, vaccination against K88 *E. coli* and other serotypes causing

PWD have been relatively unsuccessful. Fairbrother *et al.* (2005) reviewed a large body of literature and indicated that although vaccination against specific serotypes can be successful when used in field situations, efficacy is often disappointing. As noted before in this review in relation to egg yolk antibody, the likely reason for this lies in the heterogeneity of the K88 antigen. Serotyping of K88 is based on a polyclonal antibody which has a narrow range of specificities. A much more extensive knowledge about antigen heterogeneity of K88 and other serotypes in the PWD complex are required. As noted recently by Nagy & Fekete (2005) "…..our knowledge about the etiology and pathogenesis of PWD, about the existing virulence factors, about intestinal secretory activity, cellular immunity, and receptor biology is limited, and further improvements in this area are needed" (Nagy & Fekete, 2005).

2.2.5 Transgenic plants

ETEC K88 fimbriae attach to intestinal receptors found in jejunal Peyer's patches (JPP) to induce protective innate immunity stimulating antibody secreting cells (ASC) (Snoeck *et al.*, 2006). The adhesins of F4, or K88 fimbriae, are made up of repetitive protein subunits called FaeG plus smaller subunits which are responsible for adhesion to enterocytes in the small intestines. These purified fimbrial subunits are able to induce protective immunity when administered orally, although this mechanism is not well understood leading to situations in which adaptive immunity does not fully develop (Van den Broeck *et al.*, 1999). In a recent study, the genes encoding were transformed into alfalfa (*Medicago sativa* L.) plants so that transgenic plants were produced which expressed the recombinant protein. When ingested in the diet, a F4 specific immune response was induced that led to an adaptive immune response which resulted in reduced *E. coli* K88$^+$ adherence to intestinal tissue when pigs were challenged with the pathogen (Joensuu *et al.*, 2006).

Jacobs *et al.* (2006) demonstrated that colicin E7 and its immunity gene could be expressed in maize by transgenesis, and could reduce *E. coli* O157:H7 carriage in cattle if fed as a dietary supplement. A similar approach could also be used to express colicins that inhibit *E. coli* K88$^+$ in feedstuffs fed to nursery pigs. This approach is technically feasible but consumer resistance to genetically modified foods and feed is likely to hinder the use of this technology.

2.2.6 Probiotics

Recently, probiotics have gained popularity in fields of biotherapeutics and functional foods for humans and animals. Many probiotic species have been extensively studied for their role in maintaining or improving gastrointestinal health. Exact mechanisms of action of probiotic bacteria are in most cases are not known but there are at least a few examples where specific mechanisms have been identified (Fairbrother *et al.*, 2005). The concept of a probiotic is that non- pathogenic bacteria with specific properties (the probiotic) when included as a prophylactic in the diet will out compete the pathogens by a variety of mechanisms and help in disease prevention.

Probiotics were originally defined as "........microorganisms promoting the growth of other microorganisms" but the present day definition for probiotic is "mono or mixed cultures of live microorganisms which when applied to animal or man, beneficially affect the host by improving properties of the indigenous microflora" (Holzapfel *et al.*, 2001). According to this definition many microbial species can be considered probiotics, however most studies have focused on lactic acid bacteria (LAB) like *Bifidobacterium* and *Enterococcus faecalis* as probiotics against *E. coli* K88$^+$ in pigs (Holzapfel *et al.*, 2001).

Jin *et al.* (2000) evaluated the ability of *E. faecium* strain 18C23 to inhibit adhesion of *E. coli* K88$^+$ to the mucus of the small intestines of pigs. They observed as much as 90% inhibition of adhesion of *E. coli* K88$^+$ when 10^9 CFU ml^{-1} or more live *E. faecium* cells were added simultaneously with *E. coli* to immobilized mucus. Recently, Scharek *et al.* (2005) demonstrated that supplementation of the probiotic strain *E. faecium* SF68 in the feed of pregnant sows and piglets had no immune-stimulatory effect but reduced the number of pathogenic *E. coli* including β-hemolytic and *E. coli* O141 load in suckling piglets. A similar study using *Lactobacillus fermentum* strain 104R of porcine origin also inhibited adhesion of K88ab and K88ac fimbriae to ileal mucus by interacting with mucus components (Blomberg *et al.*, 1993a). It was shown that a proteinaceous substance from the spent culture of *L. fermentum* that affects mucus attachment resulted in inhibition of attachment of K88 fimbriae (Blomberg *et al.*, 1993a). Ouwehand and Conway (1996) found a proteinaceous substance of 1,700 kDa from dead *L. fermentum* strain 104r and in spent culture inhibited the adhesion of *E. coli* K88$^+$ but showed no interference with *E. coli* K88$^+$ fimbrial receptors.

A number of studies have demonstrated the role of microbes in improving non-immunologic intestinal defenses as well as improvement of innate and acquired immune responses and suppression of hypersensitivity reactions (Perdigon *et al.*, 2003; Vinderola *et al.*, 2004). Intestinal epithelium expresses major histocompatibility complex class I and II molecules that are involved in adaptive immune recognition of pathogenic organisms (Quereshi & Medzhitov., 2003). The microflora is also known to regulate intestinal inflammatory mediators like IL-10, IL-2 and IL-4 (Braat *et al.*, 2004; Pena *et al.*, 2005). Revajova *et al.* (2000) found *L. salivarius* to be more effective in enhancing immune response than *L. casei* by increasing CD2, CD4 and CD8 positive lymphocytes in peripheral blood in gnotobiotic piglets in response to a challenge by *E. coli* K88$^+$.

Lahtinen *et al.* (2005) compared four methods of testing the viability of *Bifidobacterium* strains in fermented oat products during storage. They found that subpopulations of viable but nonculturable cells retained functional cell membranes indicating dormancy of *Bifidobacteria* during storage. These dormant probiotic cells may not be useful in providing any benefits to the host. As discussed earlier, cell wall components of *Lactobacillus* spp. that are active against certain pathogenic organisms are released only when the parent bacterium is dead (Ouwehand & Conway, 1996). Therefore, even though most widely used as feed supplements, these probiotic species are not widely used as biotherapeutic agents in post-weaning diarrhea.

Kim *et al.* (2005) studied the effects of porcine-derived mucosal competitive exclusion cultures dosed at birth or weaning on growth performance and antimicrobial resistance of commensal gut *E. coli*. Piglets showed improved feed efficiency, but significantly higher numbers of commensal *E. coli* from treated piglets were resistant to tetracycline and streptomycin. Estienne *et al.* (2005) found no beneficial effects of antibiotics (erythromycin, penicillin G procaine, tylosin or oxytetracycline) given intra-muscularly or *Lactobacillus* spp. and *Streptococcus* spp. probiotics in the pre-weaning period. However, they observed enhanced average daily gain and feed consumption when the probiotic fed piglets were weaned into pens with piglets from mixed litters.

2.2.6.1 *E. coli* as a probiotic

E. coli are part of the normal flora in mammals. Moreover, *E. coli* can be acid tolerant (Bearson *et al.*, 1997; Ganzle *et al.*, 1999), and can survive acidic conditions in the stomach. Studies have used colicin producing *E. coli* to reduce pathogenic *E. coli* load in calves (Tkalcic *et al.*, 2003) however, only a few studies have investigated *E. coli* as a probiotic for swine. Schroeder *et al.* (2006) used probiotic *E. coli* strain Nissle 1917 (EcN) to prevent acute secretory diarrhea induced by enterotoxigenic *E. coli* Abbotstown (EcA) in pigs. EcN was effective in protecting animals that were given EcN as a pretreatment against EcA. EcN is well known for prevention and treatment of various diseases of the human digestive tract like Crohn's disease and ulcerative colitis (Kruis *et al.*, 1997; Malchow, 1997 Lodinova-Zadnikova & Sonnenborn, 1997). It was found to be as effective as mesalazine in maintaining remission in patients with ulcerative colitis and inhibited adhesion to intestinal epithelial cells of pathogenic *E. coli* strains isolated from patients with Crohn's disease (Kruis *et al.*, 1997; Malchow, 1997).

One of the proposed mechanisms of action of probiotic strains is by producing antimicrobial compounds called bacteriocins (Booth *et al.*, 1977; Jack *et al.*, 1995). *E. coli* strains that produce bacteriocins can be useful as biotherapeutic tools as they can potentially establish themselves in the gut microflora. Portrait *et al.* (1999) reported that poultry-derived strains of *E. coli* producing microcin J25 were successful in inhibiting *Salmonella enteritidis*. *S. enteritidis* is a causative agent of enteritis in humans.

2.2.7 Prebiotics

Diet has a significant influence on the digestive health of animals and humans. Prebiotics are fermentable carbohydrates that selectively promote growth of beneficial bacteria which in turn can stimulate digestive health of the host (Isolauri *et al.*, 2002). A probiotic used with a prebiotic that enhances the effect of the probiotic is called a synbiotic (Isolauri *et al.*, 2002). Many prebiotics are carbohydrate substrates that are not digested by the mammalian host but can be fermented in the hindgut by microorganisms. Fructooligosaccharides (FOS) are not digested in the upper gastrointestinal tract (GIT) and facilitate the passage of probiotic organisms through the upper GIT by providing them with a substrate for growth and attachment (Gibson & Roberfroid, 1995). The effect of diet and its components can readily be seen on gastrointestinal bacterial communities and diversity (Apajalahti *et al.*, 2001). In a recent study, large intestinal crypt depth was higher in FOS fed piglets than controls (Tsukahara *et al.*, 2003). Previously Jin *et al.* (1994) reported that a high straw diet in growing pigs could increase the length of villi and depth of crypts in the jejunum and ileum and increase cell proliferation in the large intestine when compared to a straw-free diet.

The products of fermentation are short chain fatty acids (SCFAs) such as acetate, butyrate, and propionate, and other products like lactate and carbon dioxide. SCFAs especially n-butyrate, serve as major energy source for the enterocytes and may stimulate intestinal cell proliferation and absorption of minerals and water. Increased butyrate concentrations have been shown to stimulate mucin production and have a trophic effect on the proximal colon of gnotobiotic rats (Meslin *et al.*, 2002). Mentschel and Claus (2003) demonstrated that butyrate inhibits apoptosis of colonocytes *in vivo*.

It has also been shown that feeding growing pigs raw potato starch can result in increased colon length (Martinez-Puig *et al.*, 2002). Raw potato starch is a type II resistant starch and is not digested by mammalian amylases because of its crystalline structure. These pigs were shown to have higher purine concentrations in the colonic digesta and a greater SCFA concentration in the proximal colon when compared to corn starch fed pigs (Martinez-Puig *et al.*, 2002).

The fact that very few *E. coli* utilize starch or inulin, so that the inclusion of these two substrates in the diet will tend to have a suppressive effect on these populations and this has been demonstrated in various studies (Tzortzis *et al.*, 2005, Bird *et al.*, 2007).

Moreover, inulin has been known to suppress the hemolytic activity of *E. coli* and can also stimulate production of colicins (Valyshev *et al.*, 2000). Inulin type fructans also modulate gut microbiota by stimulation of gut associated bifidobacteria and reduce pathogens (Kleessen & Blaut, 2005). The latter workers also demonstrated the proliferation of colon crypts and an increase in mucins from goblet cells in rats given dietary supplementation of inulin-type fructans. In contrast, guar gum and carboxy methyl cellulose, both of which are soluble and fermentable, can exacerbate the diarrhea caused by *E. coli* and *Brachyspira pilosicoli* and cause reduced weight gain in piglets (McDonald *et al.*, 1999; Hopwood *et al.*, 2002). Certain end-products of protein fermentation like branched chain fatty acids have been correlated with the up-regulation of pro-inflammatory cytokines (Pie *et al.*, 2006).

Sanderson (2004) proposed that SCFA's regulate chemokine expression in intestinal epithelial cells by inhibiting histone deacetylase activity. Chemokines are responsible for recruitment of neutrophils and lymphocytes into the intestines. Therefore increasing butyrate in the intestines can increase gut immune function. Pie *et al.* (2006) also observed up-regulation of pro-inflammatory cytokine IL-6 in the colon of carbohydrate fed weaned piglets.

2.3 *E. coli* bacteriocins - colicins

Gut microbes in the human and animal digestive tract interact extensively with the host epithelium through metabolic product exchange, which in part leads to a dynamic equilibrium (Nicholson *et al.*, 2005). Many microorganisms produce inhibitory components (antimicrobials) to reduce competition for space and nutrients in the gut. Antimicrobials produced by microorganisms include toxins, enzymes, phage viruses, metabolic byproducts, antibiotics and bacteriocins (Riley, 1998).

For at least 80 years, there have been extensive studies on different groups of molecules that are produced by bacteria as defense substances, and bacteriocins are among the best studied group of these molecules. Bacteriocins are biologically active protein molecules with a bactericidal mode of action (Lakey *et al.*, 1994). They differ from antibiotics in their narrow spectrum of activity. Colicins produced by *E. coli* are the most extensively studied bacteriocins (Riley & Wertz, 2002). Production of colicins by *E. coli* strains against other *E. coli* and related bacteria is called colicinogeny. Natural *E. coli* populations very frequently contain plasmid-encoded colicin gene clusters (colicin plasmid) (Riley & Gordon, 1992). Many strains of *E. coli* naturally occur in the GIT of human beings and animals. Some of them are known as being opportunistic pathogens, using their fimbriae and toxins as their main virulence factors. Many pathogenic *E. coli* also produce colicins, however, the role of colicins in the disease process is uncertain.

There are more than twenty colicins that can inhibit other *E. coli* and closely related bacteria. All the colicins are encoded by plasmids and show certain common features. A colicin producing *E. coli* is invariably resistant to the action of its own colicin. Colicin sensitive bacteria have receptors on their cell surface which combine with colicins to produce the lethal effects. Receptors are defined by a gene mutation that leads to resistance to one or more colicins (Alonso *et al.*, 2000). Bacteria carrying colicin plasmids (*col*) have lower growth rates (Pagie & Hogeweg, 1999). The mechanism of action of different colicins varies due to differences in ion channel formation, differences in nuclease (enzymatic) activity, and different levels of inhibition of murein synthesis in the bacterial cell wall (Lakey *et al.*, 1994). Colicins have served as an evolutionary model in studying bacteriocin development (Riley, 1998; Pagie & Hogeweg, 1999; Riley & Wertz, 2002). Riley and Wertz, 2002 focused on phylogenetic relationships among different colicins. Though there were no genetic similarities between pore forming and nuclease colicins, subgroups were created within these two groups based on genetic similarity.

2.3.1 Colicin gene

The colicin gene cluster (Fig.A-2.1) is comprised of a colicin gene, an immunity gene and a lysis gene. The colicin gene encodes for the toxic colicin, the immunity gene confers resistance to the colicin (to prevent the producer from killing itself) and the lysis gene product is involved in the release of colicin from the bacterial cell (Riley & Wertz, 2002). During stationary phase growth, when nutrients for bacterial multiplication are exhausted, organisms produce antimicrobial substances to reduce competition for scarce resources (Kleanthous & walker, 2001). Similarly, colicin genes are expressed under stress conditions and are operated by the SOS regulon, a regulatory circuit that controls genes involved in repairing damaged DNA (James *et al.*, 2002). In culture media, colicinogeny can be induced by mutagens like mitomycin C, UV radiation (Lakey *et al.*, 1994) and low doses of ciprofloxacin (Jerman *et al.,* 2005) as well as oligosaccharides like inulin (Valyshev *et al.,* 2000). Translocation is important in the ability of colicins to act on sensitive bacteria (Pugsley, 1984).

2.3.2 Classification of colicins

Based on the mechanisms of action there are two major classes of colicins, pore forming colicins and nuclease colicins. Pore or channel forming colicins (colicins A, B, E1, Ia, Ib, K, N) depolarize the inner membrane of bacterial cell walls by an ionophore like action and kill the sensitive cell by dissipating ion gradients (Alonso *et al.,* 2000). Nuclease colicins target DNA (in the case of colicins E2, E7, E8 and E9), rRNA (in the case of colicins E3, E4 and E6) or tRNA (in the case of colicins E5 and D) and digest the target DNA or RNA to make it non functional (Housden *et al.,* 2005). As mentioned earlier more than twenty colicins have been classified based on mechanism of action.

Riley (1998) focused on the evolution of different colicins and their diversification over evolutionary time (Fig.A-2.2, Appendix-A). Sequences of these proteins can be divided into two groups, the pore formers and nuclease colicins and very low levels of sequence similarities are found between these families (Fig.A-2.2, Appendix-A). Comparison of the C-terminal of these peptides indicates that there are three subfamilies within the pore formers (colicins Ia, A and E1) and two subfamilies within the nuclease colicins. In nature, selection pressure and recombination events in different colicin groups appear to be responsible for the diversification of colicins (Riley, 1998). Alternatively, colicins can be manipulated to create new colicin types. Qiu *et al.,* (2005) have engineered new anti-enterococcal peptides by the fusion of colicin Ia and pheromone cCF10 from *Enterococcus faecalis.* These peptides were shown to be highly effective against vancomycin resistant *E. faecalis* in mice (Qui *et al.,* 2005).

2.3.3 General structure of colicins

Pore forming and nuclease colicins differ in size and amino acid composition, containing 449 to 629 and 178 to 777 amino acids, respectively (Riley & Wertz, 2002). There are three domains in the colicin protein (Fig.A-2.3, Appendix-A). Translocation is carried out by the N terminus domain (T). Receptor recognition and binding to the receptor on target cells is a function of the central (R) domain, while the C terminus domain (C) is responsible for cytotoxic activity and activation of immunity (Gökce & Lakey, 2003).

Different receptors participate in the uptake of colicins into the target cells before inhibitory action of the colicin can be initiated. Gouaux (1997) has investigated the molecular basis for biological activity of the pore forming colicins, including their sequence organization, architecture, and mechanism of action. For colicins E1, Ia and A the domains for pore formation, receptor binding, and translocation of the peptide reside on a single polypeptide chain as seen in fig.A-2.4, Appendix-A. Different colicins in this group use different receptors on the outer membrane of the bacterial cell (Gouaux, 1997).

2.3.4 Colicins and pathogenic *E. coli*

The use of colicin producing *E. coli* to inhibit pathogens is not new. Braude and Siemienski (1968) were able to demonstrate that uropathogenic *E. coli* could be inhibited by a strain of *E. coli* that produced colicin V. They also demonstrated transfer of an antibiotic resistance gene spontaneously between these two *E. coli* strains. As early as 1929 commercial products appeared and Nissle (1929), colicin X (1961) and colibacterin (1973) were produced for treating bacillary dysentery in humans (Hardy, 1975).

Djonne (1985) showed that there was a higher degree of colicin resistance in enteropathogenic strains and proposed that colicins may have a role in establishing infections by enteropathogenic *E. coli* in newborn piglets. Djonne (1985) tested 315 *E. coli* isolates from piglets that died from either neonatal *E. coli* diarrhea or septicemia caused by *E. coli* and found that all enterotoxin positive *E. coli* produced colicins. Ninety nine percent of *E. coli* strains possessing the K antigen were colicinogenic. Djonne (1986) also observed a higher degree of colicin resistance in enteropathogenic *E. coli* and the majority of *E. coli* strains isolated from the intestines of piglets were resistant to colicins E1, E2, E3, Ia, and H. Guimaraes de Brito *et al.* (1998) reported that 38.7% of uropathogenic *E. coli* isolated from pigs were colicinogenic. Nemeth *et al.* (1994) found that the *E. coli* isolated from the mammary gland in bovine mastitis and from fecal samples were essentially the same and were opportunistic pathogens. Blanco *et al.* (1997) reported colicinogeny in 80% of highly pathogenic strains of *E. coli* isolated from chickens. Murinda *et al.* (1996) evaluated the inhibitory activity of colicins against *E. coli* serotypes O157,H-, O26,(H11,H-), and O111,(H8,H11,H-) and found them to be effective in the majority of cases. Jordi *et al.* (2001) found five shiga toxin producing *E. coli* (STEC) including *E. coli* O157:H7 to be sensitive to colicins E1, E4, E8-J, K and S4.

The cattle rumen is considered a major reservoir for STEC. *E. coli* O157:H7 is a virulent food borne pathogen of humans and can spread through consumption of contaminated food, milk, and is a natural inhabitant of the GIT of cattle. Schamberger and Diez-Gonzalez (2002) and Schamberger *et al.* (2004) successfully used colicin producing isolates from humans and nine different animal species to reduce fecal shedding of *E. coli* O157:H7 in cattle and also found colicin E7 to be effective against *E. coli* O157:H7. Colicins G, H, E2 and V were shown to inhibit *E. coli* O157:H7 (Bradley *et al.,* 1991).

Nandiwada *et al.* (2004) used isolated human fecal strains of *E. coli* to inhibit O157:H7 and characterized a new colicin of about 61.3 kDa which was named Hu194. This colicin belonged to the colicin E family and closely resembled colicins E2 and E7. They also evaluated the efficacy of Hu194 in controlling 22 strains of *E. coli* O157:H7 in contaminated alfalfa seeds. Alfalfa sprouts are used as salad and contaminated sprouts can transmit *E. coli* O157:H7 to humans from manure-treated alfalfa farms. Tracka and Smarda (2003) found a higher incidence of colicinogeny in enterotoxigenic *E. coli* (ETEC) than commensal *E. coli* isolated from diarrheic piglets. They also concluded that colicinogeny was not a causative factor of post weaning diarrhoea in piglets and found an abundance of colicinogenic *E. coli* in recovering piglets. Stahl *et al.*

(2004) reported *in vitro* efficacies of colicins E1 and N against *E. coli* K88$^+$ but work on the use of colicins to control this pathogen is limited.

2.3.5 Colicins summary

Colicins are peptides produced by *E. coli* that inhibit other *E. coli* and closely related bacteria and are often associated with pathogenicity of the bacteria. However, carefully selected colicinogenic strains have successfully reduced certain pathogenic *E. coli* in previous studies. Many alternatives to antibiotics have been investigated but probiotics have received particular attention because of their health and environmental benefits.

Little research has been done to investigate the effect of colicins in farm animal gastrointestinal diseases particularly infections in pigs caused by *E. coli*. Few studies have utilized colicinogenic *E. coli* as a competitive exclusion system for swine, even though colicins of *E. coli* have been shown to protect animals from gastrointestinal diseases. More research in the field of animal sciences is required to show the benefits of the natural phenomenon of colicinogeny.

3.0 Hypothesis and Objectives

In this thesis it was hypothesized that an *E. coli* probiotic could be developed that produced colicins against *E. coli* K88$^+$ but did not possess toxins harmful to pigs. The probiotic should express fimbriae so that attachment to intestinal mucosa would occur. It was further hypothesized that to enhance its competitive advantage colicinogenic strains able to utilize starch and inulin, sugars which are not usually used by *E. coli*, could be selected.

The objectives of this thesis were:

(i)   to identify colicins effective against pathogenic strains of *E. coli* bearing K88 type fimbriae;

(ii)   to select ETEC K88 specific colicinogenic *E. coli* from environmental isolates;

(iii) to characterize the identified isolates

(iv)   to identify probiotic specific carbohydrates to serve as prebiotics and

(v)   to evaluate the synbiosis of probiotics and prebiotics against *E. coli* K88$^+$ in an *in vitro* competition model.

4.0 Manuscript

Development of a Colicinogenic *Escherichia coli* K88[+] Specific Probiotic Strains for Post-weaning Piglets

4.1 Abstract

Aim of this study was to select environmental $E.$ $coli$ isolates that produced colicins against the swine pathogen $E.$ $coli$ K88$^+$. In initial evaluation using a modified plate method with 18 colicinogenic $E.$ $coli$ constructs, colicins E3, E4, E5, E9, Ia, K and N were found to possess inhibitory activity against 12 ETEC K88$^+$ strains. A total of 463 environmental isolates from cattle rumen, cattle feces, pig feces and hog manure-amended soil were screened for colicin production by a modified plate test. Further, colicinogenic isolates were screened for five toxin genes LT, STa, STb, VT1 and VT2 as well as K88 (F4) fimbriae using PCR reactions. Fourteen non-pathogenic isolates were subjected to characterization of colicin genes by PCR using 9 new primer sequences, antibiotic susceptibilities and substrate utilization. Two potential probiotic strains of $E.$ $coli$, UM-2 and UM-7 which produced colicins that could utilize potato starch and inulin were selected for in-$vitro$ competition with $E.$ $coli$ K88$^+$ strain 2-12. In $vitro$ competition between the synbiotics and $E.$ $coli$ K88$^+$ revealed inhibition of $E.$ $coli$ K88$^+$. Based on the present in $vitro$ studies it could be concluded that carefully selected potential synbiotics should be further studied for their role in protecting piglets from post-weaning diarrhea without antibiotics.

4.2 Introduction

Enterotoxigenic *E.coli* (ETEC) is a common causative agent for diarrheal diseases in both humans and animals (Padhye & Doyle, 1991; Nagy & Fekete, 2005). Post-weaning diarrhea caused by ETEC bearing K88 (F4) type fimbriae is a significant problem in the swine industry and is responsible for high mortality in newly weaned piglets (Fairbrother *et al.,* 2005). Antibiotics have been the control measure of choice for many decades in piglet diets. However their excessive use may have contributed to the occurence of multiple drug resistant pathogens in human clinical medicine and the continued use of antibiotics as feed additives is no longer acceptable to the public or governments. Thus, replacement for feed antibiotics in swine diets is needed.

Many approaches have been tested to control post-weaning diarrhea, including the immunization of sows with K88 antigen (Nagy *et al.,* 1985) and dietary inclusion of a range of compounds including zinc oxide, organic acids, probiotics, prebiotics, spray dried porcine plasma (Owusu-Asiedu *et al.,* 2003; Bosi *et al.,* 2004; Yi *et al.,* 2005) and egg yolk antibodies (Owusu-Asiedu *et al.,* 2003; Chernysheva *et al.,* 2003). Probiotics are receiving great interest as feed supplements but many of the claims made by manufacturers are not always backed up by scientifically verifiable data. In the last few years, this situation has changed with the use of *E. coli* strain Nissle and *Lactobacillus* VSL in the treatment of inflammatory bowel disease, the efficacy of which was demonstrated in human clinical trials (Kruis *et al.,* 1997; Malchow *et al.,* 1997; Chapman *et al.,* 2007).

Recently a series (Schamberger & Diez-Gonzalez, 2002; Schamberger & Diez-Gonzalez 2004; Schamberger *et al.,* 2004) of publications described the development of a colicinogenic *E. coli* strain that inhibited the colonization of *E. coli* O157:H7 in cattle. The probiotic was selected from a range of environmental isolates and produced colicin E7. When inoculated into cattle experimentally infected with O157:H7 the abundance of the pathogen was reduced by approximately one log in the feces. Similar positive results have been obtained by inoculating urinary catheters with colicin producing strains of *E. coli* (Trautner *et al.,* 2005). Stahl *et al.* (2004) extracted a range of colicins from *E. coli* and demonstrated *in vitro* that the growth of *E. coli* K88[+] was inhibited in the presence of colicins E1 and N.

The present study was undertaken to select nonpathogenic environmental *E. coli* isolates as a probiotic for swine production. In this study, a suite of environmental isolates of *E. coli* obtained from pig manure, cattle rumen content, cattle feces, and soil were screened for their ability of inhibit the growth of *E. coli* K88[+] obtained from clinical specimens of piglet post-weaning diarrhea. Selected isolates were partially characterized for colicin type, antibiotic susceptibilitites and substrate utilization profiles. Selected isolates were found effective in reducing the growth of K88[+] positive strains of *E. coli* and their efficacy *in vitro* was demonstrated.

4.3 Methods (See Fig.B-4.2, Appendix-B for experimental approach)

Bacterial strains utilized. Eighteen reference strains of *E. coli* K-12 constructs with plasmids encoding the Pugsley (1984) reference colicins (Table 4.1) were provided by Dr Francisco Diez-Gonzalez (University of Minnesota, MN, USA). Twelve clinical *E. coli* K88[+] strains isolated from nursery pigs were obtained from Dr Carlton Gyles (University of Guelph, ON, Can., Table 4.1). A total of 463 strains of *E. coli* isolated from a variety of ecosystems were from our own laboratory collection. All standard bacterial stock cultures were stored at -80°C in sterile 50% v/v glycerol. Strains were routinely cultured in Luria Bertani (LB) broth or Luria Bertani (LBa) agar (Fisher) at 37°C and regularly streaked out to single colonies to ensure purity.

Detection of colicin activity. The procedure for detection of colicins was a modification of the procedure described by Schamberger and Diez-Gonzalez (2002). The sensitivity of 12 K88 reference strains was tested against the colicin producing *E. coli* constructs. Well separated colonies were picked and inoculated into LB broth and grown overnight (12-16 h). Each overnight culture (100 µl) of K88 (indicator strain) was streaked on LB agar plates containing 0.025 mg.L^{-1} of mitomycin C (LBm) (Acros Organics, NJ, USA) and allowed to dry (Fig.B-4.1, Appendix-B). Mitomycin C induces the production of colicins. From each of 18 *E. coli* K12 reference constructs (test strains), 7 µl of overnight culture was spotted onto the LBm plates previously streaked with the indicator K88 strains and allowed to dry. These plates were then incubated at 37°C for 12 to 16 h. Thus a total of 216 (18 test strains x 12 K88 indicator strains) individual plate assays were performed. Zones of clearing (see fig.B-4.4, Appendix- B) around the test strains were measured with an image analysis device and digital calipers (Alpha Innotech, San Leandro, CA, USA). A minimum measurement of 1mm for zone of inhibition was considered as inhibitory to *E. coli* K88$^+$. The most sensitive K88 reference isolate was strain 2-12 (Table 4.1) and it was used as the indicator strain for screening of the 463 environmental isolates for colicin production. All 463 environmental isolates were tested in the plate inhibition assay against *E. coli* K88$^+$ strain 2-12 at least in duplicates, and positive assays showing inhibition were repeated again. Limitation of plate method cannot be ignored as other antimicrobial compounds can also be induced by mitomycin C. Putative colicin like activity using the above mentioned test will be refered to as colicinogeny for the current study.

DNA isolation and PCR amplification. Bacterial cultures were spun down at 3,500 x *g* for 10 min, and the supernatant was discarded. The cell pellet was resuspended in 400 μl TE buffer (10 mM Tris-HCl, 1 mM Na$_2$EDTA), again centrifuged (3,500 x *g*), the supernatant discarded, the pellet resuspended in 400 μl TE buffer, and heated at 95°C for 15 min to lyse the cells and release the DNA. To purify the DNA, 400 μl of phenol:chloroform:isoamylalchohol (25:24:1) was added to the samples and vortexed for 30 sec. This mixture was incubated at -70°C for 15 min and then centrifuged at 10,000 x *g* (Microlite, Thermo IEC, Waltham, MA, USA). Approximately 300 μl of supernatant was transferred to a 1.5 ml microfuge tube, 300 μl of chloroform was added, mixed and centrifuged at 10,000 x *g*. A supernatant sample of 200 μl was stored at -20°C for future PCR analysis.

PCR reactions. PCR reactions were in a 25 μl volume and comprised 1 μl of MgCl$_2$ (15 mM), 0.5 μl of 2.5 mM dNTP, 2.5 μl of *Taq* buffer (10x), 0.5 μl of each primer (25 pM each), 0.1 μl of *Taq* DNA polymerase (New England Biolabs,USA), and 2 μl of the DNA sample. Sterile filtered water was used to bring the final reaction volume to 25 μl. The *E. coli* K-12 colicin producing constructs for 18 different classes of colicins were used as colicinogenic references (Table 4.1), and the presence of colicin genes was confirmed by PCR analysis using newly designed primers (Table 4.2) from GenBank sequences (Fig.C-4.1 to C-4.10, Appendix-C). The thermocycler (Techne Genius, Duxford, Cambridge, UK) cycling conditions were: one cycle of denaturation at 94°C for 2 min, annealing at primer specific Tm measured by using equation 1 (G+C) +2(A+T) number or else optimized, (Table 4.2) for 1 min, elongation at 72°C for 1 min; then 35 cycles of denaturation at 94°C for 1 min, annealing at primer specific Tm (Table 4.2) for 1 min, elongation at 72°C for 1 min and then a final hold temperature of 72°C for 5 min. PCR reaction products were stored at -20°C or immediately separated on 2% agarose gels in a 0.5X TBE buffer (10X TBE is 108 g Tris base, 55 g boric acid, and 40 ml of 0.5M Na$_2$EDTA). Samples were electrophoresed on a 2% agarose gel containing 0.5μg/ml ethidium bromide (Digital images of gels were taken using an image analysis device (Alpha Innotech). See fig.B-4.5 Appendix-B for representative pictures of agarose gel electrophoretogram.

Screening for toxin genes. All colicin positive environmental strains (121 isolates) were tested for the presence of toxin genes LTp, STa STb, VT1 and VT2 and the K88 fimbriae gene (Fig 4.4). DNA was extracted and PCR reactions using specific primers were carried out (Table 4.2) by above mentioned method. The 14 most colicinogenic (largest zone of clearing around test strain) isolates were selected for detailed genotyping. Results for zones of inhibitions are present in Fig. 4.3.

Antibiotic sensitivity test. Antibiotic sensitivity testing was done according to the National Center for Clinical and National Laboratory Standards (NCCLS, 2002) procedures. Isolates to be tested were streaked on LB agar plates, and well separated colonies were inoculated into Mueller-Hinton broth (MH) and allowed to grow for 6 h at 35°C. The turbidity of the inoculum used was adjusted to a 0.5 McFarland standard. A McFarland standard of 0.5 is achieved by mixing 0.05 mL of a 1.175 % $BaCl_2.2H_2O$ solution, with 9.95 mL of 1% H_2SO_4. A sterile cotton swab was dipped in the culture and the entire agar surface of a MH agar plate was streaked three times. Antibiotic disks were dispensed on the surface of the agar after it was dry and plates were incubated for 16 h at 37°C. Digital images were taken with an image analysis device (Alpha Innotech), and zones of clearing around the disks were measured with pixel calipers provided by the Alpha Innotech software. Only 14 selected potentially probiotic strains and five randomly selected K88[+] strains (Table 4.1) were subjected to antimicrobial susceptibility assays two times described above and results are present in Table 4.5.

Substrate utilization profiles. Substrate utilization patterns of *E. coli* isolates were determined using API 50 CH (bioMerieux, St Laurent, QC, Can). See Table B-4.1 for list of substrates used in API 50 CH. Pure cultures from isolated colonies were used for the tests as described above. A colony was picked with a sterile swab and resuspended in one ml of sterile water to give a McFarlane optical density of approximately four and added to 10 ml of API 50 CHB/E medium (bioMerieux). A 120 µL volume of cell rich medium was dispensed into each test strip capsule containing a test substrate. Two incubation period readings were taken. One, at 24 ± 2 h and the second at 48 ± 6 h. A positive test was indicated by a change in color of media from red to yellow for most carbohydrates, and from red to brown for esculin and ferric citrate. A negative reaction was red or orange for most carbohydrates, red to magenta for potassium gluconate and potassium 2-keto-gluconate (Fig.B-4.6, Appendix-B) Only 14 selected potentially probiotic environmental isolates (Table 4.4) and five randomly selected K88$^+$ strains (Table 4.1) were subjected to the same substrate fermentation tests two times as described above and results for *E. coli* K88$^+$ are present in Table 4.5.

***In vitro* competition assays.** Isolates *E. coli* UM-2 and UM-7 (Table 4.4) from cattle feces were evaluated by *in vitro* competition assays with *E. coli* K88$^+$ strain 2-12. UM-2 and UM-7 were made resistant to levofloxacin by repeated transfer in LB broth containing 1 µg. ml^{-1} of levofloxacin. The MIC for both was determined to be 0.05 µg. ml^{-1}. *E. coli* 2-12 was sensitive to this concentration of levofloxacin. *E. coli* 2-12 was resistant to ciprofloxacin at 0.05 µg. ml^{-1} and resistance was increased to 4 µg. ml^{-1} by repeated transfers in LB with increasing concentrations of ciprofloxacin. Individual strains were maintained on minimal medium which contained (g. L^{-1}): glucose, 5; Na$_2$HPO$_4$, 6; KH$_2$PO$_4$, 3; NH$_4$Cl, 1; MgSO$_4$, 0.12; CaCl$_2$, 0.01. For competition assays strains were transferred at least three times on minimal medium with the growth substrate to be assayed. For example, for the competition assay with starch, the glucose in the minimal medium was substituted with starch. 2% minimal media were used by adding 2g either of starch or inulin in 100 ml of minimal medium. The *E. coli K88*[+] 2-12 strain plus UM-2 or UM-7 (mixed culture) was inoculated into medium to give a final cell concentration of approximately 10^6 cfu ml[-1] and incubated at 37°C. Semi-batch type culture testing was performed by inoculating 100μl of mixed culture from 12 h growths to 10 ml of sterile minimal medium or LB tubes. This was carried out from 0h to 36 h at 12 h intervals. At the end of each incubation period, one ml of a well mixed culture was serially diluted in triplicate in buffered peptone water and plated onto LB agar and Eosine-methylene blue (EMB) agar with or without 4 μg. ml[-1] ciprofloxacin as well as levofloxacin at 37°C for 16 h. All competition experiments were repeated on three separate occasions.

Statistical analysis. All data were compiled and analyzed with Fischer's t-test using JMP 5.1 (Sall *et al.*, 2001).

4.4 Results

Among *E. coli* K-12 colicin producing constructs for 18 different colicins used as colicinogenic references (Table 4.1) the presence of 14 colicins classes was clearly determined by PCR analysis using 9 primers (Table 4.2) designed for the current study, but E3, E9, Ia, and Ib were ambiguous. *E. coli* K88[+] indicator strains from clinical cases of piglet post-weaning diarrhea were positive for F4 fimbriae and heat labile (LT) toxin genes (Table 4.1). As expected, none of the *E. coli* K-12 constructs that produced colicins were positive for the K88 gene or LT gene. All 463 environmental isolates were tested in the plate inhibition assay against *E. coli* K88$^+$ strain 2-12 at least in duplicate and all positive isolates produced repeatable results.

The 463 *E. coli* isolates were obtained from cattle feces (358), rumen fluid (33), swine feces (35), and soil (37) (Table 4.3). Of these strains 96/358 from cattle feces, 0/33 from rumen fluid, 9/35 from swine feces, and 16/37 from soil inhibited the growth of indicator strain 2-12. Thus, 121/463 environmental strains were colicin positive and 71/121 were found negative for toxins genes (LT, STa, STb, VT1 and VT2) and selected 14/71 toxin negative colicinogenic strains for further screening (Fig 4.3).

The 14 environmental strains producing the largest zone of inhibition (Fig.4.3) around *E. coli* K88$^+$ strains 2-12 were subjected to further detailed analysis including of carbohydrate utilization test, antibiotic sensitivity assay, colicin classification using PCR reactions, confirmation of absence of toxins genes and hemolysis on blood agar (Table 4.4). Primers (Table 4.2) differentiated the 14 different colicins produced by the *E. coli* K-12 constructs into 9 groups (Table 4.2). Each group-specific PCR had a different size amplicon (Table 4.2) and primers from one group did not amplify sequence from a non-target group organism. There were three strains (UM-3, 5, and 11 in Table 4.3) that appeared to have a novel colicin because the primers based on the Pugsley (1984) strains did not produce an amplicon.

The 14 most colicinogenic environmental strains were tested for their ability to ferment a range of substrates (Table 4.4). Table 4.4 only represents the differing substrate utilization patterns and these were amygdalin, arbutin, D-cellobiose, esculin, gentiobiose, inositol, L-sorbose, methyl-α-D-glycopyranoside, and salacin. Only two strains, UM-2 and UM-7, were able to utilize starch and inulin. These strains were highly sensitive to ciprofloxacin, less so to imipenem, but resistant to ampicillin, erythromycin, kanamycin, oxytetracycline, penicillin, streptomycin, and tetracycline. No zone of inhibition was obtained for vancomycin which was the negative control.

Five randomly selected K88^{+} strains (Table 4.1) were subjected to the same substrate fermentation tests and antimicrobial inhibition assays described above and results are present in Table 4.5. The differing substrates for the pathogenic strains were adonitol, arabinose, sorbitol, esculin, and gentiobiose. These strains could not utilize inulin or starch. Antimicrobial resistance profiles were similar to those of the environmental isolates.

UM-2 and UM-7 grew in broth culture on starch and inulin and both strains were passaged at least 15 times without any diminution of growth characteristics. In competition assays with UM-2 (Fig. 4.1) or UM-7 (Fig. 4.2) the probiotic strains were significantly ($P < 0.05$) more abundant than *E. coli* 2-12. These assays resulted in the same overall results when repeated on three different occasions with the approximately same bacterial counts for start culture and at same culture conditions. Different starches (wheat, rice, and corn) were also tested but the results were similar to those of the potato starch.

4.5 Discussion

Post-weaning diarrhea and edema disease are some of the most common diseases found in weaned pigs in North America and Europe (Fairbrother & Nadeau, 2006). In a recent case-control study in the Canadian Province of Ontario consisting of 28 case nurseries and 22 control nurseries, more than 60% of symptomatic nurseries were positive for $E.$ $coli$ K88$^+$ (Amezcua et $al.$, 2002). Significant numbers of case positive farms reported mortality of 20-30% over a one to two month period. Even in low mortality (5-10%) nurseries the economic loss (based on financial conditions at the time) in a 500-sow herd would have amounted to $20,000 US annually. Given the demands to remove sub-therapeutic antibiotics from nursery pig diets because of the potential risk to human health, alternative strategies to controlling post-weaning diarrhea and edema disease must continually evolve.

One approach to developing therapeutics to $E.$ $coli$ K88$^+$ is probiotics. Probiotics theoretically have metabolic activities like the production of antimicrobial compounds, selective substrate utilization patterns, or immunogenic activity (Marco et $al.$, 2006), that individually or collectively, competitively exclude a pathogen from its preferred niche in the gut ecosystem (Riley & Wertz, 2002). In the current study, the approach used was to inhibit $E.$ $coli$ producing the F4 enterocyte adhesin (K88) factor by colicin antimicrobial activity. Colicin production is an allelopathic mechanism that relies on the production of a short antimicrobial peptide by *E. coli* against other members of the same species (Riley & Wertz, 2002). There are currently at least 9 groups (Riley & Wertz, 2002), and with associate sub-groups we screened for 18 different colicins (Table 4.1). In the current study, it was found that only colicins E3, E4, E5, E9, K, N, and Ia producing *E. coli* K-12 constructs inhibited all 12 K88 clinical strains screened (Table 4.1). All strains of K88 tested were inhibited by these seven colicins but no other work has been done on the assessment of different colicinogenic *E. coli* against K88 and this makes it difficult to generate comparisions with the present data.

Stahl *et al.* (2004) screened the ability of purified colicin E1 and N to inhibit K88, but did not evaluate other colicins. In contrast, more work has been done on enterohaemorrhagic *E. coli* O157:H7. Schamberger *et al* (2002) screened a range of environmental isolates (540) of *E. coli* against O157:H7 and only colicin E7 was effective against all O157:H7 strains tested. Subsequent animal trials with experimentally dosed O157:H7 demonstrated that about a 2 log reduction in O157:H7 could be obtained with the *E. coli* colicinogenic strain. Zhao *et al.* (1998) inoculated a cocktail of O157:H7 inhibiting bacteria consisting of different bacterial strains, some of which were *E. coli*, into cattle experimentally infected with O157:H7 and found that there was a reduction in shedding of the pathogen.

The present data thus appear to be unique as far as K88 is concerned, because this is the first report to our knowledge that has demonstrated a range of colicins that are all active against K88. Schamberger and Diez-Gonzalez (2002) only found one colicin active against 22 *E. coli* O157:H7 strains, but this result may be a consequence of strain variation. In the present work, approximately 25% of 463 environmental isolates produced a colicin against the K88 indicator strain (2-12). Schamberger and Diez – Gonzalez (2002) screened 540 environmental strains against a universal indicator strain based on the common laboratory *E. coli* K-12. Only those that inhibited the growth of K-12 were subsequently screened against O157:H7. It is possible that the initial screening against K-12 by the latter authors selected for a subset of environmental strains that resulted in reducing the range of colicin types that were potentially active against O157:H7. In other words screening against K-12 was based on a narrow definition of genotypic susceptibility while the present tests were based on broad genotype variation.

An interesting aspect of the present screening approach was that all colicinogenic strains of *E. coli* were isolated from non-swine environments (Table 4.3). For example, active strains were isolated from cattle feces but not from the rumen of cattle. Based on anti-*E. coli* K88$^+$ activity, this would imply that the rumen *E. coli* population is different from that in the hind gut represented by feces. Schamberger and Diez – Gonzalez (2002) screened isolates of *E. coli* from the feces of cats, cattle, chickens, deer, dogs, ducks, horses, humans, pigs, and sheep. They found colicinogenic strains from all sources but the highest number was from cats and sheep. No rumen isolates were evaluated. Thus, it was not unexpected that colicinogenic strains could be isolated from the diverse environmental sources tested here.

When the sugar fermentation profiles of the colicinogenic environmental isolates were compared they differed in substrate specificity (Table 4.4). The K88 strains differed from the environmental strains primarily in their ability to ferment sorbose, and arabinose. Two of the environmental isolates (UM-7 and UM–2) obtained from cattle feces were able to ferment inulin and starch which appears to be an unusual characteristic of *E. coli* (Schultz & Strockbine, 2005). Since utilization of these carbohydrates is an unusual feature for *E. coli*, their use in hog diets may provide a unique opportunity for synbiotic therapy.

Probiotics are microbial cultures that have a biological effect on the host due to production of fermentation end-products and antimicrobial molecules, and which may modulate immune function (Tuohy *et al.*, 2005; Bengmark, 2003). On the other hand, prebiotics are fermentable sugars that modify gut fermentation so that microbial populations considered beneficial to gut health, like *Lactobacillus spp.* proliferate (Tuohy *et al.*, 2005; Bengmark, 2003). Synbiotics are a combination of the two; microbial cultures which in their own right have a beneficial effect, but in the presence of a fermentable sugar enhanced beneficial gut effects can be elicited (Tuohy *et al.*, 2005; Bengmark, 2003).

Present studies clearly indicated that UM-2 and UM-7 out-competed *E. coli* K88 (Fig 4.1 & 4.2). However, from the present study, it is not possible to determine the mode of their inhibitory action. When LB was used as the growth medium, UM-2 and UM-7 out-competed K88, but because both K88 and UM strains can grow equally well alone on LB it is likely that the colicins have a role in the competition. In the case of starch and inulin, because K88 does not utilize these substrates, but the UM strains do, the extent of the inhibitory contribution of the colicins to the overall inhibition noted is unclear. The synbiotic effect is based on the fact that very few *E. coli* apparently utilize starch or inulin so that the inclusion of these two substrates in the diet will tend to have a suppressive effect on these populations as has been demonstrated in other studies (Tzortzis *et al.*, 2005; Bird *et al.*, 2007). Secondly, because the UM strains produce group N/S4 colicins they should be able to compete effectively for enterocyte adhesion sites with pathogenic *E. coli* K88[+].

Surprisingly, a review of studies that have investigated the use of synbiotic therapy to combat *E. coli* K88[+] infections is not available. However, there are a number of studies that link the use of prebiotics like inulin, fructooligosaccharides, and other oligosaccharides to gut health in pigs (Flickinger *et al.*, 2003). For weaned pigs there is also a significant volume of literature on the use of probiotics (Stein & Kil, 2006), but little of this is related specifically to *E. coli* K88[+]. Branner and Roth-Maier (2006) used the prebiotics lactulose, inulin, and mannanoligosaccharides in combination with an *Enterococcus faecium* probiotic to examine their effects on the availability of B-vitamins. Shim *et al.* (2005) examined the feeding of an antibiotic-free creep feed supplemented with either oligofructose, probiotics or synbiotics to suckling piglets. Although they did not use *E. coli K88[+]* as a response criterion they concluded that oligofructose or synbiotics added to the antibiotic-free creep feed during the pre-weaning period benefit gut microbial populations and performance of piglets.

In conclusion, a large collection of environmental isolates of *E. coli* were found able to inhibit the growth of *E. coli* K88$^+$ clinical strains. Two of these strains grew well on starch and inulin, two substrates that are frequently used as prebiotics in weanling pig diets. *In vitro* assays demonstrated that these two potentially probiotic strains out-competed K88, but their mode of inhibitory action was not characterized. The utility of these potentially probiotic strains must be tested in an *in vivo* model of *E. coli* diarrhea to determine whether the *in vitro* effects carry over to live animals.

4.6 Acknowledgements

This study was financially supported by the Manitoba Pork Council and the Agri-Food Research & Development Initiative of the Province of Manitoba. We gratefully acknowledge Roman Kotlowski for contributing to primer designs and excellent technical assistance. The authors thank Dr. C. L. Gyles (University of Guelph) and Dr. F. Diez-Gonzalez (University of Minnesota) for providing the reference *E. coli* strains.

4.8 Tables and Figures

Table 4.1 List of reference strains used in this study and their characteristics

Reference E. coli strains	Colicin	Clinical strain	F4(K88)	LT	Inhibitory to K88$^+$(2-12)	Reference
BZB2101	Colicin A	No	-	-	No	Pugsley , 1984[*]
BZB2102	Colicin B	No	-	-	No	Pugsley, 1984
BZB2104	Colicin E1	No	-	-	No	Pugsley, 1984
BZB2125	Colicin E2	No	-	-	No	Pugsley, 1984
BZB2106	Colicin E3	No	-	-	Yes	Pugsley, 1984
BZB2107	Colicin E4	No	-	-	Yes	Pugsley, 1984
BZB2108	Colicin E5	No	-	-	Yes	Pugsley, 1984
BZB2150	Colicin E6	No	-	-	No	Pugsley, 1984
BZB2110	Colicin E7	No	-	-	No	Pugsley, 1984
Pap1407	Colicin E9	No	-	-	Yes	Pugsley, 1984
BZB2119	Colicin Ia	No	-	-	Yes	Pugsley, 1984
BZB2115	Colicin Ib	No	-	-	No	Pugsley, 1984
BZB2103	Colicin D	No	-	-	No	Pugsley, 1984
BZB2116	Colicin K	No	-	-	Yes	Pugsley, 1984
Pap 1	Colicin M	No	-	-	No	Pugsley, 1984
BZB2123	Colicin N	No	-	-	Yes	Pugsley, 1984
Pap 1401	Colicin V	No	-	-	No	Pugsley, 1984
2-11	E.coliK88$^+$	K88/pig	+	+	NA	Gyles C L[†]
2-12	E.coli K88$^+$	K88/pig	+	+	NA	Gyles C L
JG-306	E.coli K88$^+$	K88/pig	+	+	NA	Gyles C L
JG-280	E.coli K88$^+$	K88/pig	+	+	NA	Gyles C L
DAKH	E.coli K88$^+$	K88/pig	+	+	NA	Gyles C L
DAKA	E.coli K88$^+$	K88/pig	+	+	NA	Gyles C L
A1	E.coli K88$^+$	K88/pig	+	+	NA	Gyles C L
1-27	E.coli K88$^+$	K88/pig	+	+	NA	Gyles C L
1-37	E.coli K88$^+$	K88/pig	+	+	NA	Gyles C L
1-74	E.coli K88$^+$	K88/pig	+	+	NA	Gyles C L
PDH-8	E.coli K88$^+$	K88/pig	+	+	NA	Gyles C L
1-36	E.coli K88$^+$	K88/pig	+	+	NA	Gyles C L

+ Present; - Absent; LT- Heat labile toxin; NA- not applicable as not tested against E. coli K88$^+$.

*See reference Pugsley (1984)

†All E. coli K88$^+$ strains were kind gifts from Dr. C L Gyles, University of Guelph, Canada.

Table 4.2 PCR primers used in this study.

Target gene	Primer name	Primer sequences (5' to 3')	PCR product size [bp]	Annealing temp °C	Reference
Colicin A,N, S4	NS4f NS4r	CGTAGCTATAATGAAGCAATGGCTTCA ACCTCCAACAGGAGAGGTCCCCAGTT	225	57	This study*
Colicin M	Mf Mr	CCAGCAACCCTCTCACATTGCAG CCAGAAAACATCGCCCCGAGCC	556	68	This study
Colicin V	Vf Vr	CACGCCCTGAAGCACCACCA CCGTTTTCCAAGCGGACCCC	400	68	This study
Colicin Ia, Ib	Iabf Iabr	GCACAACAGGCCCGTCTGCTC CACCTTCCACATCCTCTGTCACC	385	68	This study
Colicin E2, E3, E3a, E4, E5, E6, E7, E8, E9	Mixf Mixr	CGACAGGCTAAAGCTGTTCAGGT TGCAGCAGCATCAAATGCAGCCT	219	60	This study
Colicin U, Y	YUf YUr	GTGAACGGACAGAAACCCGCC CAATCTGTCTGACAGCCTCTCCC	243	68	This study
Colicin B, D, D157	BDf BDr	TCGCTCCATCCATGCCTCCG CCATCCCGACCAGTCTCCCTC	138	68	This study
Colicin 5, 10, K	510Kf 510Kr	AAAGCTGAACTGGCGAAGGC CAACTCATCATCCCCTATGTAAGAAG	803	60	This study
Colicin E1	E1f E1r	ACGGGAGTGGCTCTGGCGG CTCTTTACGTCGTTGTTCTGCTTCCTG	389	68	This study
Fimbrae K88	K88f K88r	GCACATGCCTGGATGACTGGTG CGTCCGCAGAAGTAACCCCACCT	439	68	This study
Heat stable toxin STa	STaf Star	GTGAAACAACATGACGGGAGG ATAACATCCAGCACAGGCAGG	251	60	Kotlowski *et al* (2006)
Heat stable toxin STb	STbf STbr	GGGGTTAGAGATGGTACTGCTGGAG GACAATGTCCGTCTTGCGTTAGGAC	181	68	Kotlowski *et al* (2006)
rRNA 16S	27f 342r	GAAGAGTTTGATCATGGCTCAG CTGCTGCCTCCCGTAG	352	55	Kotlowski *et al* (2006)
Verotoxin VT1	VT1f VT1r	CGCATAGTGGAACCTCACTGACGC CATCCCCGTACGACTGATCCC	91	65	Kotlowski *et al* (2006)
Verotoxin VT2	VT2f VT2r	CGGAATGCAAATCAGTCGTCACTCAC TCCCCGATACTCCGGAAGCAC	265	65	Kotlowski *et al* (2006)
Heat labile toxin LTp	LTf LTr	CCGTGCTGACTCTAGACCCCCA CCTGCTAATCTGTAACCATCCTCTGC	480	68	Kotlowski *et al* (2006)

*Primers for this study were designed by using GenBank sequences and using Clustal software for multiple alignments (Benson *et al.*, 2006; Wheeler *et al.*, 2006; Hall, 1999).

Table 4.3 Sources of environmental *E. coli* isolates and distribution of colicinogeny among isolates.

Source of *E. coli* Isolates	Number (%) of Isolates	Colicinogenic (%) inhibitory to K88	Non-colicinogenic (%)
Cattle Feces	358(77.3)	96(26.8)	262(73.1)
Cattle Rumen Fluid	33(7.1)	0(0)	33(100)
Swine Feces	35(7.5)	9(25.7)	26(74.28)
Hog Manure amended soil	37(7.9)	16(43.2)	21(56.7)
Total	463 (100)	121(26.13)	342(73.9)

E. coli from these sources were tested for colicin production using plate test for this study.

Table 4.4 Phenotypic characterization of the most colicinogenic environmental *E. coli* isolates.

Test	UM-1	UM-2	UM-3	UM-4	UM-5	UM-6	UM-7	UM-8	UM-9	UM-10	UM-11	UM-12	UM-13	UM-14
								Strains						
Carbohydrate Fermentation profile*														
Amygdalin	-	+	-	-	-	-	-	-	+	-	-	-	-	-
Arabutin	-	+	-	-	-	-	-	-	+	-	-	-	-	-
D-Cellobiose	-	+	-	-	-	-	-	-	+	-	-	-	-	-
Esculin + ferric citrate	+	+	+	+	-	+	+	+	+	+	+	+	+	+
Gentiobiose	+	+	+	-	+	+	+	+	+	+	+	+	+	+
Inositol	-	+	-	-	-	-	+	+	+	+	+	+	+	+
Inulin	-	+	-	-	-	-	+	-	-	-	-	-	-	-
L-Sorbose	-	-	-	-	-	-	-	-	+	-	-	-	-	-
Methyl- α D-Glucopyranoside	-	+	-	-	-	-	-	-	-	-	-	-	-	-
Salicin	-	-	-	-	-	-	-	-	+	-	-	-	-	-
Starch	-	+	+	-	-	-	+	-	-	-	-	-	-	-
Antibiotic Susceptibility Test†														
Zone of Inhibition (mm)														
Antibiotic (MIC)														
Ampicillin (≥32)	6.31	6.96	7.94	7.91	6.04	6.98	6.16	2.21	6.81	6.00	5.70	8.23	8.21	9.03
Ciprofloxacin (1->4)	13.79	16.8	14.77	17.22	14.61	13.22	15.69	11.93	16.64	15.35	17.22	17.44	16.33	16.98
Imipenem (8-≥16)	7.06	12.4	13.35	13.84	11.27	9.55	11.40	13.05	13.90	12.63	12.44	11.33	13.76	13.26
Kanamycin (≥64)	4.58	8.9	7.77	7.00	6.39	4.04	6.77	7.91	6.89	6.78	8.29	9.17	9.63	9.46
Oxytetracyclin (≥16)	9.25	10.95	9.25	8.26	9.78	8.16	9.22	7.85	9.39	7.95	8.07	6.60	8.40	8.24
Streptomycin (Na)	3.69	4.12	3.83	4.62	3.5	2.12	4.13	4.03	3.63	3.67	6.11	7.71	3.74	4.38
Tetracyclin (≥16)	7.99	9.14	8.84	9.53	9.37	8.78	9.88	8.57	10.10	7.40	8.48	7.4	8.62	8.31
Vancomycin (NC)	0	0	0	0	0	0	0	0	0	0	0	0	0	0
Colicin Typing‡	B/D	N/S4, B/D	Na	K/5/10	Na	N/S4,V, B/D	N/S4	N/S4,E1, K/5/10	K/5/10	B/D	Na	Group E	K/5/10, B/D	K/5/10 B/D
Enterotoxins§														
LTp, STb	-	-	-	-	-	-	-	-	-	-	-	-	-	-
STa, VT1 or VT2	-	-	-	-	-	-	-	-	-	-	-	-	-	-

NC- Negative control; Na- Not available.; - Negative reaction/absence

*All selected environmental *E. coli* tested positive for Indole, Methyl Red test, Glycerol , D-Arabinose, L-Arabinose, D-Ribose, D-Xylose, D-Galactose, D-Glucose, D-Fructose, D-Manose, L-Rhamnose, Dulcitol, D-Manitol, D-Sorbitol, N-Acetyl glucosamine, D-Maltose, D-Lactose (Bovine origin), D-Melibiose, D-Saccharose, D-Trehalose, D-Raffinose, L-Fucose and Potassium Gluconate. All strain including reference strains of *E. coli* K88⁺ showed negative reaction for Voges-Proskauer, Citrate utilization, Erythritol, l-Xylose, D-Adonitol,Methyl-β D-Xylopyranoside, D-Fucose, D-Arabitol, L-Arabitol, Potassium-2-ketogluconate, Potassium-5-ketoguconate,D-Turanose, D-Lyxose, D-Tagatose, Xylose, Glycogen, Melezitose, D-Cellobiose, Methyl-α D-Mannopyranoside.

†Antibiotic sensitivity tests were carried out using Disk Diffusion test performed according to NCCLS (2002)

‡ Colicins determined using PCR. Colicin N sequence confirmed using RFLP using *Dde*-I and *Mse*-I restriction enzymes Group E consists of

Table 4.5 Phenotypic characterization of *E. coli* K88[+] strains

Test	Strain				
Carbohydrate Fermentation profile*	**K88 2-11**	**K88 2-12**	**K88 1-36**	**K88 J-280**	**K88 P-8**
D-Adonitol	-	-	-	+	-
D-Arabinose	+	+	+	-	-
D-Sorbitol	+	+	+	+	-
Esculin + ferric citrate	-	+	-	-	-
Gentiobiose	-	-	-	-	+
Antibiotic Susceptibility Test[†]	**Zone of Inhibition (mm)**				
Ampicillin-10 µg ($\geq$32)	7.20	6.30	7.70	0	0
Ciprofloxacin-5µg (1.5-$\geq$4)	14.29	13.60	10.05	16.46	14.43
Erythromycin-15 µg ($\geq$8)	5.66	2.77	1.33	3.29	0
Imipenem-10 µg (8-$\geq$16)	12.93	12.84	11.04	12.66	10.35
Kanamycin-30 µg ($\geq$64)	10.14	10.94	8.46	0	0
Oxytetracyclin-30 µg ($\geq$16)	8.57	8.35	0	0	0
Streptomycin-10 µg (na)	0	1.19	4.37	0	0
Tetracyclin-30 µg ($\geq$16)	10.14	8.18	0	0	0
Vancomycin-30 µg (NC)	0	0	0	0	0
Enterotoxins[§]					
LTp, STb	+	+	+	+	+
STa, VT1 orVT2	-	-	-	-	-

NC-Negative control ; na - not available; + Positive reaction/ presence; - Negative reaction/absence

*All reference strains of *E.coli* K88[+] tested positive for Indole test, Methyl Red test Glycerol, L- Arabinose, D-Ribose, D-Xylose, D-Galactose, D-Glucose, D-Fructose, D-Manose, L-Sorbose, L-Rhamnose, Dulcitol, D-Sorbitol, N-AcetylGlucosamine, D-Maltose, D-Lactose, D-Melibiose, D-Saccharose (Sucrose) ,D-Trehalose, D-Raffinose, L-Fucose, Potassium Gluconate, , Hemolysis on 5% blood agar, and negative for Erythritol, L-Xylose, Methyl-βD-Xylopyranoside, Inositol, Methyl-αD-Mannopyranoside, Methyl- αD-Glucopyranoside, Amygdalin, Arbutin, Salicin, D-Cellobiose, Inulin, D-Melezitose, Starch, Glycogen, Xylitol, D-Turanose, D-Lyxose, D-Tagatose, D-Fucose, D-Arabitol, L-Arabitol, Potassium 2-KetoGluconate, Potassium 5-KetoGluconate, Voges-Proskauer test, Citrate utilization test.

[†]Antibiotic sensitivity test carried out using Disk Diffusion test

[§]Enterotoxins LTp, STa, STb, VT1 and VT2 determined using PCR.

Figure legends

Fig. 4.1 *In vitro* competition of *E. coli* UM-2 with *E. coli* K88 strain 2-12 in (a) LB, (b) starch (2%), or (c) inulin (2%) based medium at 37°C with shaking at 150 rpm. Viable numbers from two independent experiments are reported. * indicates significant difference in counts of UM-2 and *E. coli* K88$^+$ 2-12.

Fig. 4.2 *In vitro* competition of *E. coli* UM-7 with *E. coli* K88 strain 2-12 in (a) LB, (b) starch (2%), or (c) inulin (2%) based medium at 37°C with shaking at 150 rpm. Viable numbers from two independent experiments are reported. * indicates significant difference in counts of UM-2 and *E. coli* K88$^+$ 2-12.

Fig. 4.3 Zones of inhibition (mm) are averages of four independent observations measured using calipers. Zones are produced by a colicinogenic *E. coli* when spotted on lawn of *E.coli* K88$^+$ strain 2-12.

Fig. 4.4 Cluster analysis of enterotoxin genes from colicinogenic environmental strains brings more of soil and cattle fecal isolates together. 50 colicin producing isolates were tested positive for one or more toxins at a given time.

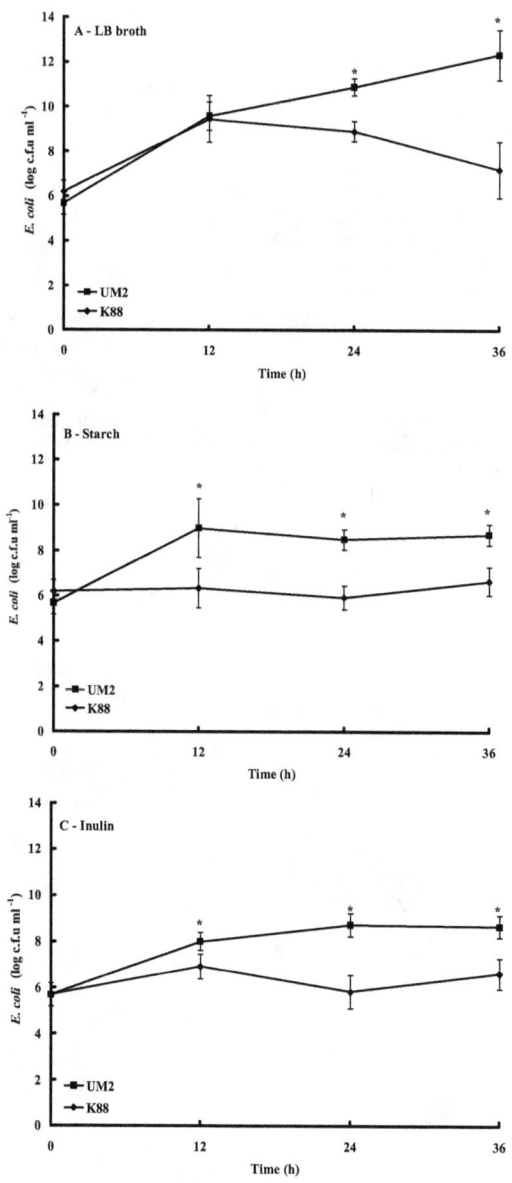

Figure 4.1

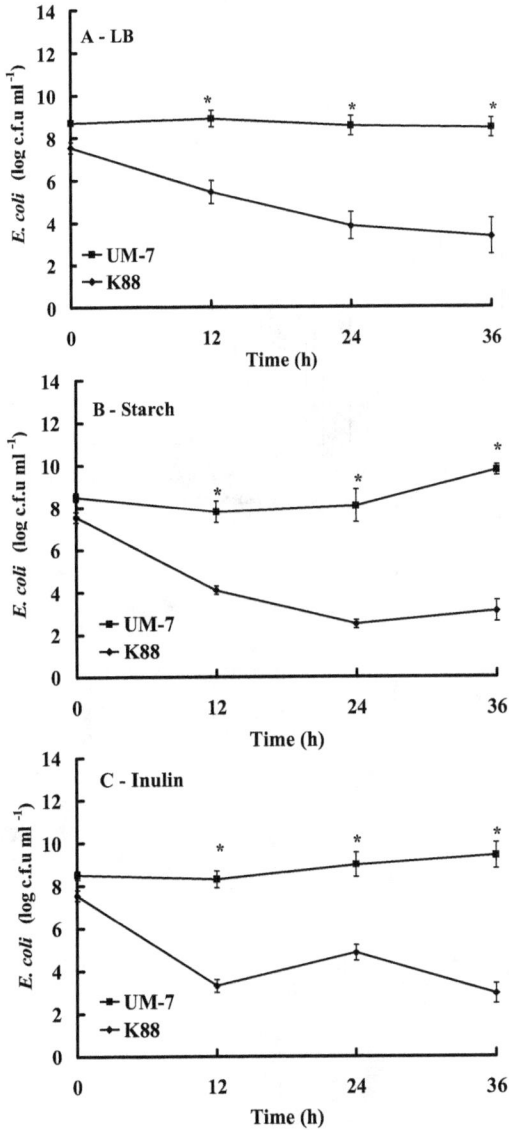

Figure 4.2

Fig. 4.3 Averages of zones of inhibitions of top 14 isolates from selected potential probiotic *E. coli*

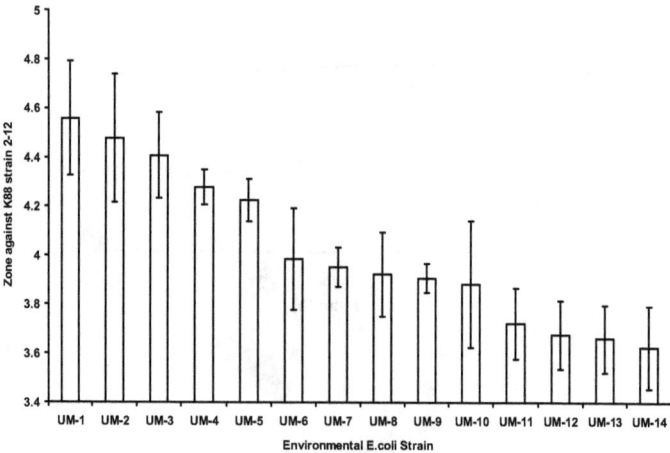

Fig. 4.4 Cluster analysis of enterotoxins from colicinogenic environmental strains.

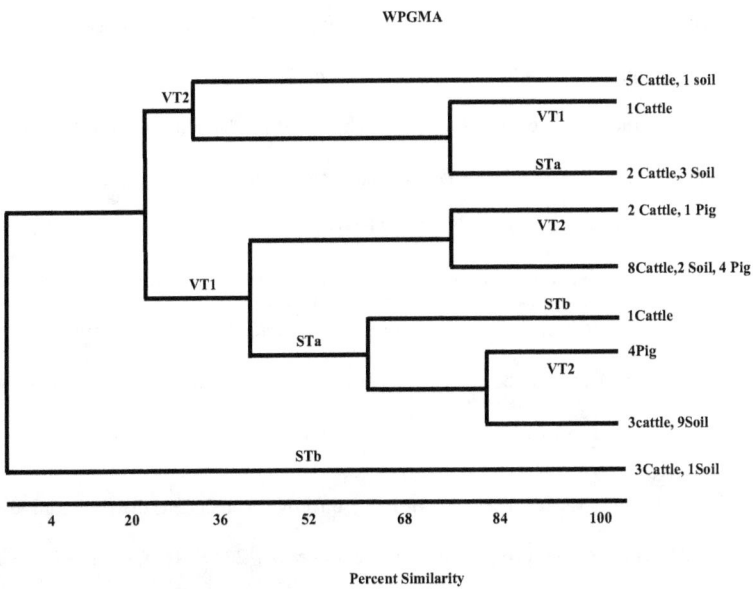

5.0 General Summary

Increasing antimicrobial resistance in pathogens and increasing antibiotic residues in environment is leading to the demand to remove antibiotics as sub-therapeutic feed supplements in swine production (Amezcua *et al.*, 2002; Nagy & Fekete, 2005). Post-weaning diarrhea and the cost associated with loss of production often present a major problem to farms wishing to practice antibiotic free swine production (Amezcua *et al.*, 2002; van Beers-Schreurs *et al.*, 1992). A better strategy is required to successfully wean pigs without feeding antibiotic additives. Many alternatives to antibiotics have been investigated but probiotics have received particular attention because of their general acceptance as being safe to host and environment (Walker & Duffy, 1998).

The use of probiotics has been shown to be effective in reducing diarrhea caused by ETEC K88 strains (Blomberg *et al.*, 1993a; Jin *et al.*, 2000; Revajova *et al.*, 2000). However, the consistency of results has always been questionable (Nagy & Fekete, 2005). Traditionally, Gram-positive lactic acid bacteria have been used as probiotic strains (Holzapfel *et al.*, 2001). Although use of Gram-negative species as a probiotic is rare in swine production, there have been recent studies on the use of fungi, yeasts and some Gram-negative microorganisms as probiotics (Isolauri *et al.*, 2002; Schroeder *et al.*, 2006). Other studies have confirmed the use of *E. coli* as probiotics in human conditions (Kruis *et al.*, 1997; Malchow, 1997; Lodinova-Zadnikova & Sonnenborn, 1997). One of the best examples is *E. coli* strain Nissle 1917 which is commonly used as a probiotic to treat Crohn's disease and ulcerative colitis in humans (Kruis *et al.*, 1997; Malchow, 1997).

In the current study, it was shown that potentially probiotic strains of *E. coli* UM-7 and UM-2 were inhibitory and very effective in competing with *E. coli* K88$^+$ in an *in vitro* system, generating results which are consistent with other results in the literature.

In this study, screened 463 strains of *E. coli* isolated from soil, pig and cattle feces for their ability to produce putative colicin activity against clinical indicator strains of *E. coli* K88$^+$ obtained from Dr. Carlton Gyles (University of Guelph). Colicins effective against *E. coli* K88$^+$ were tested using 18 reference colicins which were encoded on plasmid pBR322 in the common *E. coli* K-12 host. The absence of toxin genes for heat stable (STa and STb), heat labile (LT) and vero (VT1 and VT2) toxin production by colicin producers was confirmed by PCR reactions.

- Approximately 26% of strains from cattle and pig feces produced putative colicins against *E. coli* K88$^+$, but more than 43% of strains from soil inhibited *E. coli* K88$^+$. From the plate inhibition test it was concluded that colicin E3, E4, E5, E9, N, K and Ia inhibited all 12 *E. coli* K88$^+$ tested.

- Of the 121 strains isolated with colicin like activity against *E. coli* K88$^+$, 50 isolates possessed one or more of the 5 toxin genes tested by PCR reactions. 32 isolates were positive for VT1, 7 for VT2, 23 for STa, 3 for STb genes and none for LT gene.

- Fourteen toxin gene negative isolates were selected based on maximum zone of inhibition around *E. coli* K88$^+$ on plate test for puatative colicin production, for utilization of carbohydrate sources from panel of 49 different carbohydrates. Two of these isolates UM-2 and UM-7 utilized starch and inulin as sole carbon sources, whereas none of five tested *E. coli* K88$^+$ utilized starch or inulin as sole carbon source.

- It was demonstrated that *in vitro* competition between *E. coli* K88$^+$ strain 2-12 and two putative colicinogenic isolates UM-2 and UM-7 resulted in significant inhibition of *E. coli* K88$^+$ in the presence either of 2% starch or 2% inulin in minimal media as also seen in standard LB medium at 37°C in semi-batch culture .

Results from this *in vitro* study are promising because selection of strains was based on colicin production as a mechanism of action of these potentially probiotic *E. coli* strains. It is suggested that selected strains may have use as a probiotic for weaned piglets, although much more data from animal trials are required before conclusions can be drawn safely.

6.0 Conclusions and Future Research

- Based on the results from *in vitro* studies, it is recommended to use putative colicin producing *E. coli* strains UM-2 and UM-7 along with starch and or inulin (synbiotic combination). As yet, these probiotics have not been tested in an *in vivo* model. The importance of testing in an *in vivo* model cannot be overstated, as results *in vitro* do not always carry over to the *in vivo* situation. As part of the *in vivo* testing it is critical that an *E. coli* K88$^+$ induced disease model be used, as it is insufficient to rely on naturally occurring diarrhea which could be the result of other agents (eg. rotavirus), or simply a nutritional diarrhea that occurs because of poorly digested feed ingredients.

- A second issue for future studies if initial *in vivo* testing is successful is field testing. This is likely to be challenging because there are typically a large array of factors in the field including various carbohydrate sources always present in swine feeds, other than the factors studied in the laboratory that would make a synbiotic effective. In practice, a herd with a history of *E. coli* K88$^+$ diarrhea should be chosen and probiotics added to the feed of these animals. Secondly a control herd with similar clinical and management characteristics should to be recruited as the control group.

- In this study we tested strains for efficacy against a single *E. coli* K88$^+$ strain (2-12) largely because of logistical needs. The reality is that the actual antigen in K88 can be highly heterogenous. By this, it is meant that K88 is identified as a bacterium that precipitates with a particular anti-K88 antibody in an immunoprecipitation assay. The actual gene, and consequently the protein that is encoded on the fimbrae is heterogenous among strains and it is actually the structure of this antigen that

59

determines the pathogenicity of the bacterium. In the field an array of genotypically variant K88 types will be encountered, not all of which may be susceptible to the probiotic we have developed.

- More research should be carried out to find more colicins active against *E. coli* K88[+]. We only had 18 reference strains that produced known colicins, however, to date at least 25 colicins have been classified. This is likely to be a highly productive area of research as the more colicins, particulary with novel activities that can be found the larger will be the arsenal available to control disease.

- The source of the majority of our collection of colicinogenic *E. coli* was cattle feces and soil from pastures where these animals grazed. A community genetic analysis of colicinogenic isolates would explain whether there is a relationship between these *E. coli* and their persistence in the environment. By sampling from a wider range of environments it is likely that novel colicins can be found.

- The current data shows that UM-2 or UM-7 inhibit *E. coli* K88[+] on plate test and out-compete *E. coli* K88[+] in *in vitro* competition assay, not how it occurred. Similarly, in the competition assays with starch or inulin as the substrate, it is not known if a colicin was expressed or whether the dominance of the colicin producers was only the result of the inability of K88 to utilize starch or inulin. The classical bacterial genetics approach to investigate a problem of this type is to create isogenic mutants to a gene (eg. colicin gene knockout) and then see what the results in a competition assay would be.

7.0 Literature Cited

Alonso, G., Vilchiz, G. & Rodriquez Lemoine, V. (2000). How bacteria protect themselves against channel-forming colicins. *Int Microbiol* **3**, 81-88.

Amezcua, R., Friendship, R. M., Dewey, C. E., Gyles, C. & Fairbrother, J. M. (2002). Presentation of postweaning *Escherichia coli* diarrhoea in southern Ontario prevalence of hemolytic *E. coli* serogroups involved and their antimicrobial resistance patterns. *Can J Vet Res* **66**, 73-78.

Apajalahti, J. H., Kettunen, A., Bedford, M. R. & Holben, W. E. (2001). Percent G+C profiling accurately reveals diet-related differences in the gastrointestinal microbial community of broiler chickens. *Appl Environ Microbiol* **67**, 5656-5667.

Bearson, S., Bearson, B. & Foster, J. W. (1997). Acid stress responses in enterobacteria. *FEMS Microbiol Lett* **147**, 173-180.

Bengmark, S. (2003). Use of some pre-, pro- and synbiotics in critically ill patients. *Best Pract Res Clin Gastroenterol* **17**, 833-848.

Benson, D. A., Karsch-Mizrachi, I., Lipman, D. J., Ostell, J. & Wheeler, D. L. (2006). GenBank. *Nucl Acids Res* **34**, D16-D20.

Bertschinger, H. U., Nief, V. & Tschape, H. (2000). Active oral immunization of suckling piglets to prevent colonization after weaning by enterotoxigenic *Escherichia coli* with fimbriae F18. *Vet Microbiol* **71**, 255-267.

Biagi, G., Piva, A., Moschini, M., Vezzali, E. & Roth, F. X. (2006). Effect of gluconic acid on piglet growth performance, intestinal microflora, and intestinal wall morphology. *J Anim Sci* **84**, 370–378.

Biomérieux Canada (2006). Henri-BourassaWest, St. Laurent, QC.

http://www.biomerieux.com/

Bird, A. R., Vuaran, M., Brown, I. & Topping, D. L. (2007). Two high-amylose maize starches with different amounts of resistant starch vary in their effects on fermentation, tissue and digesta mass accretion, and bacterial populations in the large bowel of pigs. *Br J Nutr* **97**, 134-44.

Blanco, J. E., Blanco, M., Mora, A. & Blanco, J. (1997). Production of toxins (enterotoxins, verotoxins and necrotoxins) and colicins by *Escherichia coli* strains isolated from septicemic and healthy chicken relationship with *in vivo* pathogenicity. *J Clin Microbiol* **35**, 2953-2957.

Blomberg, L., Henriksson, A. & Conway, P. L. (1993a). Inhibition of adhesion of *Escherichia coli* K88 to piglet ileal mucus by *Lactobacillus spp. Appl Environ Microbiol* **59**, 34-39.

Blomberg, L., Krivan H. C., Cohen, P. S. & Conway, P. L. (1993b). Piglet ileal mucus contains proteins and glycolipids (galactosylceramide) receptors specific for *Escherichia coli* K88 fimbriae. *Infect Immun* **61**, 2526-2531.

Booth, S. J., Johnson J. L. & Wilkins, T. D. (1977). Bacteriocin production by strains of bacteroides isolated from human feces and the role of these strains in the bacterial ecology of the colon. *Antimicrob Agents Chemother* **11**, 718-724.

Bosi, P., Casini, L., Finamore, A., Cremokolini, C., Merialdi, G., Trevisi, P., Nobili, F. & Mengheri, E. (2004). Spray-dried plasma improves growth performance and reduces inflammatory status of weaned pigs challenged with enterotoxigenic *Escherichia coli* K88. *J Anim Sci* **82**, 1764-72.

Braat, H., van den Brande, J., van Tol, E., Hommes, D., Peppelenbosch, M. & van Deventer, S. (2004). *Lactobacillus rhamnosus* induces peripheral hyporesponsiveness in stimulated CD4+ T cells via modulation of dendritic cell function. *Am J Clin Nutr* **80**, 1618-1625.

Bradley, D. E., Howard, S. & Lior, H. (1991). Colicinogeny of O157:H7 enterohemorrhagic *Escherichia coli* and the shielding of colicin and phage receptors by their O-antigenic side chains. *Can J Microbiol* **58**, 977-983.

Bradley, R., Collee, J. G. & Liberski, P. P. (2006). Variant CJD (vCJD) and bovine spongiform encephalopathy (BSE): 10 and 20 years on: part 1. *Folia Neuropathol* **44**, 93-101.

Branner, G. R. & Roth-Maier, D. A. (2006). Influence of pre-, pro-, and synbiotics on the intestinal availability of different B-vitamins. *Arch Anim Nutr* **60**, 191-204.

Braude, A. I. & Siemienski, J. S. (1968). The influence of bacteriocins on resistance to infection by gram-negative bacteria II. Colicin action, transfer of colicinogeny and transfer of antibiotic resistance in urinary infections. *J Clin Invest* **47**, 1763-1773.

Caloca, M. J., Suárez, S. & Soler, J. (2002). Identification and partial purification of K88ab *Escherichia coli* receptor proteins in porcine brush border membranes. *Int Microbiol* **5**, 91-94.

Cassels, T. A. & Wolf, M. K. (1995). Colonization factors of diarreagenic *E. coli* and their intestinal receptors. *J Ind Microbiol* **15**, 214-226.

Celemin, C., Rubio, P., Echeverria, P. & Suarez, S. (1995). Gene toxin patterns of *Escherichia coli* isolated from diseased and healthy piglets. *Vet Microbiol* **45**, 121-7.

Chapman, T. M., Plosker, G. L. & Figgitt, D. P. (2007). Spotlight on VSL#3 probiotic mixture in chronic inflammatory bowel diseases. *Biodrugs* **21**, 61-63.

Chernysheva, L. V., Friendship, R M, Gyles, C. L. & Dewey, C. E. (2003). Field trial assessment of the efficacy of specific egg-yolk antibody product for control of postweaning *E. coli* diarrhea. *Vet Therapeutics* 4, 279-284.

Cherrington, C. A., Hinton, M., Mead, G. C. & Chopra, I. (1991). Organic acids: chemistry, antibacterial activity and practical applications. *Adv Microb Physiol* **32**, 87-108.

De Graaf, F. K. & Gaastra, W. (1994). Fimbriae of enterotoxigenic *Escherichia coli*. In: Klemm, P. (ed.) Fimbriae: Adhesion, genetics, biogenesis and vaccines. CRC press, Boca Raton, 57-88 pp.

De Graaf, F. K. & Mooi, F. R. (1986). The fimbrial adhesins of *Escherichia coli*. *Adv Microb Physiol* **28**, 65–143.

Djonne, B. K. (1985). Colicin production in relation to pathogenicity factors in strains of *Escherichia coli* isolated from the intestinal tract of piglets. *Act Vet Scand* **26**,145-152.

Djonne, B. K. (1986). Colicin resistance in relation to pathogenicity factors in strains of *Escherichia coli* isolated from the intestinal tract of piglets. *Act Vet Scand* **27**, 115-123.

Estienne, M. J., Hartsock, T. G. & Harper, A. F. (2005). Effects of antibiotics and probiotics on suckling pig and weaned pig performance. *Intern J Appl Res Vet Med* **3**, 303-308.

Fairbrother, J. M. & Nadeau, E. (2006). *Escherichia coli*: on-farm contamination of animals. *Rev Sci Tech* **25**, 555-69.

Fairbrother, J. M., Nadeau, E. & Gyles, C. L. (2005). *Escherichia coli* in postweaning diarrhea in pigs: an update on bacterial types, pathogenesis, and prevention strategies. *Anim Health Res Rev* **6**, 17-39.

Flickinger, E. A., Van Loo, J. & Fahey, G. C. Jr. (2003). Nutritional responses to the presence of inulin and oligofructose in the diets of domesticated animals: a review. *Crit Rev Food Sci Nutr* **43**, 19-60.

Francis, D. H., Grange, P. P., Zeman, D. H., Baker, D. R., Sun, R. & Erickson, A. K. (1998). Expression of mucin-type glycoprotein K88 receptors strongly correlates with piglet susceptibility to K88$^+$ enterotoxigenic *Escherichia coli*, but adhesion of this bacterium to brush borders does not. *Infect Immun* **66**, 4050-4055.

Gaastra, W. & de Graaf, F. K. (1982). Host specific fimbrial adhesins of non-invasive enterotoxigenic *Escherichia coli* strain. *Microbiol Rev* **46**, 129-161.

Ganzle, M. G., Hertel, C., van der Vossen, J. M. & Hammes, W. P. (1999). Effect of bacteriocin-producing lactobacilli on the survival of *Escherichia coli* and *Listeria* in a dynamic model of the stomach and the small intestine. *Int J Food Microbiol* **48**, 21-35.

Gibson, G. R. & Roberfroid, M. B. (1995). Dietary modulation of the human colonic microbiota: introducing the concept of prebiotics. *J Nutr* **125**, 1401-12

Gökce, I. & Lakey, J. H. (2003). Production of an *E. coli* toxin protein; colicin A in *E. coli* using an inducible system. *Turk J Chem* **27**, 323-331.

Gouaux, E. (1997). The long and short of colicin action, the molecular basis for the biological activity of channel-forming colicins. *Structure* **5**, 313-317.

GuimaraÄes de Brito, B., Leite, D. S., Linhares, R. E. C. & Vidotto, M. (1998). Virulence-associated factors of uropathogenic *Escherichia coli* strains isolated from pigs. *Vet Microbiol* **65**, 123-132.

Hall, T. A. (1999). BioEdit: a user-friendly biological sequence alignment editor and analysis program for windows 95/98/NT. *Nucl Acids Symp Ser* **41**, 95-98.

Hardy, K. G. (1975). Colicinogeny and related phenomena. *Bacteriol Rev* **39**, 464-515.

Harmsen, M. M., van Solt, C. B., Hoogendoorn, A., van Zijderveld, F. G., Niewold, T. A. & van der Meulen, J. (2005). Escherichia coli F4 fimbriae specific llama single-domain antibody fragments effectively inhibit bacterial adhesion in vitro but poorly protect against diarrhoea. *Vet Microbiol* **111**, 89-98.

Holzapfel, W. H., Haberer, P., Geisen, R., Björrkroth, J. & Schillinger, U. (2001). Taxonomy and important features of probiotic microorganisms in food and nutrition. *Am J Clin Nutr* **73** (Suppl), 365–373.

Hopwood, D. E., Pethick, D. W. & Hampson, D. J. (2002). Increasing the viscosity of the intestinal contents stimulates proliferation of enterotoxigenic *Escherichia coli* and *Brachyspira pilosicoli* in weaner pigs. *Br J Nutr* **88**, 523-532.

Housden, N. G., Loftus, S. R., Moore, G. R., James, R. & Kleanthous, C. (2005). Cell entry of enzymatic bacterial colicins, porin recruitment and the thermodynamics of receptor binding. *PNAS* **102**, 13849-13854.

Isolauri, E., Kirjavainen, P. V., Salminen, S. (2002). Probiotics: a role in the treatment of intestinal infection and inflammation? *Gut* **50** (Suppl III), 54-59.

Jack, R. W., Tagg, J. R. & Ray B. (1995). Bacteriocins of gram-positive bacteria. *Microbiol Rev* **59**, 171-200.

Jacobs, J. L., Diez-Gonzalez F. & Ronald P. (2006). Colicinogenic maize: inhibition of pathogenic *E. coli* O157:H7. In proceedings of 48th annual maize genetics conference, Pacific Grove, California. 9-12 March 2006. p 141.

James, R., Penfold, C. N., Moore, G. R. & Kleanthous, C. (2002). Killing of *E. coli* cells by E group nuclease colicins. *Biochemie* **84**, 381-389.

Jerman, B., Butala, M. & Zgur-Bertok, D. (2005). Sublethal concentrations of ciprofloxacin induce bacteriocin synthesis in *Escherichia coli*. *Antimicrob Agents Chemother* **49**, 3087-3090.

Jin, L. Z., Baidoo, S. K., Marquardt, R. R. & Frohlich, A. A. (1998). *In vitro* inhibition of adhesion of enterotoxigenic *Escherichia coli* K88 to piglet intestinal mucus by egg-yolk antibodies. *FEMS Immunol Med Microbiol* **21**, 313-321.

Jin, L. Z., Marquardt, R. R. & Zhao, X. (2000). A strain of *Enterococcus faecium* (18C23) inhibits adhesion of enterotoxigenic *Escherichia coli* K88 to porcine small intestine mucus. *Appl Environ Microbiol* **66**, 4200-4204.

Jin, L., Reynolds, L. P., Redmer, D. A., Caton, J. S. & Crenshaw, J. D. (1994). Effects of dietary fiber on intestinal growth, cell proliferation, and morphology in growing pigs. *J Anim Sci* **72**, 2270-2278.

Joensuu, J. J., Verdonck, F., Ehrstromc A., Peltola M., Siljander-Rasi H., Nuutila A. M., Oksman-Caldentey K. M., Teeri T. H., Cox E., Goddeeris, B. M. & Niklander-Teeri, V. (2006). F4 (K88) fimbrial adhesin FaeG expressed in alfalfa reduces F4+ enterotoxigenic *Escherichia coli* excretion in weaned piglets. *Vaccine* **24**, 2387-2394.

Jordi, B. J. A. M., Boutaga, K., Van Heeswijk, C. M. E., Knapen, F. V. & Lipman, L. J. A. (2001). Sensitivity of shiga toxin-producing *Escherichia coli* (STEC) strains for colicins under different experimental conditions. *FEMS Microbiol Lett* **204**, 329-334.

Kahn, C. M. (2005). (Ed.) Intestinal diseases in pigs. Enteric collibacillosis. in the Merck veterinary manual. Ninth edition. Merck & Co., N.J., USA. 246 pp.

Kanehisa, M., Goto, S., Hattori, M., Aoki-Kinoshita, K. F., Itoh, M., Kawashima, S., Katayama, T., Araki, M. & Hirakawa, M. (2006). From genomics to chemical genomics: new developments in KEGG. *Nucleic Acids Res* **34**, D354-D357. http://www.genome.jp/dbget-bin/www_bfind?drug

Kim, L. M., Gray, J. T., Bailey, J. S., Jones, R. D. & Fedorka-Cray, P. J. (2005). Effect of porcine-derived mucosal competitive exclusion culture on antimicrobial resistance in *Escherichia coli* from growing piglets. *Foodborne Pathog Dis* **2**, 317-329.

Kleanthous, C. & Walker, D. (2001). Immunity proteins: enzyme inhibitors that avoid the active site. *Trends Biochem Sci* **26**, 624-631.

Kleessen, B. & Blaut, M. (2005). Modulation of gut mucosal biofilms. *Br J Nutr* **93**, 535-540.

Kluge, H., Broz. J. & Eder, K. (2006). Effect of benzoic acid on growth performance, nutrient digestibility, nitrogen balance, gastrointestinal microflora and parameters of microbial metabolism in piglets. *J Anim Physiol Anim Nutri* **90**, 316–324.

Kotlowski, R., Bernstein, C. N., Sepehri, S. & Krause, D. O. (2006). High prevalence of *Escherichia coli* belonging to the B2+D phylogenetic group in inflammatory bowel disease. *Gut* online: http://gut.bmj.com/cgi/content/abstract/gut.2006.099796v1

Kruis, W., Schutz, E., Fric, P., Fixa, B., Judmaier, G. & Stolte, M. (1997). Double blind comparision of an oral *Escherichia coli* preparation and mesalazine in maintaining remission of ulcerative colitis. *Aliment Pharmacol Ther* 11, 853-858.

Lahtinen, S. J., Ouwehand, A. C., Reinikainen, J. P., Korpela, J. M., Sandholm, J. & Salminen, S. J. (2005). Intrinsic properties of so-called dormant probiotic bacteria, determined by flow cytometric viability assays. *Appl Environ Microbiol* 72, 5132-5134.

Lakey, J. H., Van Goot, F. G. & Pattus, F. (1994). All in a family, the toxic activity of pore forming colicins. *Toxicology* 87, 85-108.

Lodinova-Zadnikova, R. & Sonnenborn, U. (1997). Effect of preventive administration of a non-pathogenic *Escherichia coli* strain on the colonization of the intestine with microbial pathogens in newborn infants. *Biol Neonate* 71, 224-232.

Malchow, H. A. M. D. (1997). Crohn's disease and *Escherichia coli*; a new approach to maintain remission of colonic Crohn's disease? *J Clin Gastroenterol* 25, 653-658.

Marco, M. L., Pavan, S. & Kleerebezem, M. (2006). Towards understanding molecular modes of probiotic action. *Curr Opin Biotechnol* 17, 204-210.

Marquardt, R. R., Jin, L. Z., Kim, J., Fang, L., Frohlich, A. A. & Baidoo, S. K. (1999). Passive effect of egg yolk antobodies against enterotoxigenic *Escherichia coli* K88+ infection in neonatal and early-weaned piglets. *FEMS Immunol Med Microbiol* 23, 283-288.

Martinez-Puig, D., Perez, J. F., Castillo, M., Andaluz, A., Anguita, M., Morales, J. & Gasa, J. (2003). Consumption of raw potato starch increased colon length and fecal excretion of purine bases in growing pigs. *J Nutr* 133, 134-139.

McDonald, D. E., Pethick, D. W., Pluske, J. R. & Hampson, D. J. (1999). Adverse effects of soluble non-starch polysaccharide (guar gum) on piglet growth and experimental colibacillosis immediately after weaning. *Res Vet Sci* **67**, 245-250.

Mentschel, J. & Claus, R. (2003). Increased butyrate formation in the pig colon by feeding raw potato starch leads to a reduction of colonocyte apoptosis and a shift to the stem cell compartment. *Metabolism* **52**, 1400-1405.

Meslin, J. C., Bensaada, M., Popot, F. & Andrieux, C. (2001). Differential influence of butyrate concentration on proximal and distal colonic mucosa in rats born germ-free and associated with a strain of *Clostridium paraputrificum*. *Comp Biochem Physiol A Mol Integr Physiol* **128**, 379-384.

Murinda, S. E., Roberts, R. F. & Wilson, R. A. (1996). Evauation of colicins for inhibitory activity against diarrheagenic *Escherichia coli* strains, including serotype O157:H7. *Appl Environ Microbiol* **62**, 3196-3202.

Nagy, B. & Fekete, P. Z. (1999). Enterotoxigenic *Escherichia coli* (ETEC) in farm animals. *Vet Res* **30**, 259–284.

Nagy, B. & Fekete, P. Z. (2005). Enterotoxigenic *Escherichia coli* in veterinary medicine. *Int J Med Miicrobiol* **295**, 443-454.

Nagy, B., Casey, T. A. & Moon, H. W. (1990). Phenotype and genotype of *Escherichia coli* isolated from pigs with postweaning diarrhea in Hungary. *J Clin Microbiol* **28**, 651-653.

Nagy, L. K., MacKenzie T. & Painter, K. R. (1985). Protection of the nursing pig against experimentally induced enteric colibacillosis by vaccination of dams with fimbrial antigens of *E. coli* (K88, K99 and 987P). *Vet Rec* **117**, 408-413.

Nandiwada, L. S., Schamberger, G. P., Schafer, H. W., & Diez-Gonzalez, F. (2004). Characterization of an E2-type colicin and its application to treat alfalfa seeds to reduce *Escherichia coli* O157:H7. *Int J Food Microbiol* **93**, 267– 279.

Nataro, J. P. & Kaper, J. B. (1998). Diarrheagenic *Escherichia coli*. *Clin Microbiol Rev* **11**, 142-201.

National Committee for Clinical Laboratory Standards (2002). Performance standards for antimicrobial susceptibility testing; twelfth informational supplement (M100-S12), Vol. 22, No.1. Wayne, PA.

Nemeth, J., Muckle, C. A. & Gyles, C. L. (1994). *In vitro* comparison of bovine mastitis and fecal *Escherichia coli* isolates. *Vet Microbiol* **40**, 231-238.

Nicholson, J. K., Holmes, E. & Wilson, I. D. (2005). Gut microorganisms, mammalian metabolism and personalized health care. *Nat Rev Microbiol* AOP, doi,10.1038/nrmicro1152.

Noamani, B. N., Fairbrother, J. M. & Gyles, C. L. (2003). Virulence genes of O149 enterotoxigenic *Escherichia coli* from outbreaks of postweaning diarrhea in pigs. *Vet Microbiol* **97**, 87-101.

Ouwenhand, A. C. & Conway, P. L. (1996). Purification and characterization of a component produced by *Lactobacillus fermentum* that inhibits the adhesion of K88 expressing *Escherichia coli* to porcine ileal mucus. *J Appl Bacteriol* **80**, 311-318.

Owusu-Asiedu, A., Nyachoti, C. M. & Marquardt, R. R. (2003). Response of early weaned pigs to an enteropathogenic *Escherichia coli* (K88) challenge when fed diets containing spray dried porcine plasma or pea protein isolate plus egg yolk antibody, zinc oxide, fumaric acid or antibiotic. *J Anim Sci* **81**, 1790-1798.

Padhye, N. V. & Doyle, M. P. (1991). Rapid procedure for detecting enterohemorrhagic *Escherichia coli* O157:H7 in food. *Appl Environ Microbiol* **57**, 2693-2698.

Pagie, L. & Hogeweg, P. (1999). Colicin diversity, a result of eco-evolutionary dynamics. *J Theor Biol* **196**, 251-261.

Palmer, N. C. & Hulland, T. J. (1965) Factors predisposing to the development of coliform gastroenteritis in weaned pigs. *Can Vet J* **6**, 310-316.

Pedersen, C., Jonsson, H., Lindberg, J. E. & Roos, S. (2004). Microbiological characterization of wet wheat distillers' grain, with focus on isolation of lactobacilli with potential as probiotics. *Appl Environ Microbiol* **70**, 1522-1527.

Pena, J. A., Rogers, A. B., Ge, Z., Ng, V., Li, S.Y., Fox, J.G. & Versalovic, J. (2005). Probiotic *Lactobacillus spp.* diminish *Helicobacter hepaticus*-induced inflammatory bowel disease in interleukin-10-deficient mice. *Infect Immun* **73**, 912-920.

Perdigon G, Locascio M, Medici M, Pesce de Ruiz Holgado A, Oliver G.(2002). Interaction of bifidobacteria with the gut and their influence in the immune function. *Biocell* **27**, 1-9.

Pie, S., Awati, A., Vida, S., Falluel, I., Williams, B. A. & Oswald., I. P. (2006). Effects of added fermentable carbohydrates in the diet on intestinal pro-inflammatory cytokine specific mRNA content in weaned piglets. *J Anim Sci* **22**, (ahead of print).

Portrait, V., Gendron-Gaillard, S.,Cottenceau, G. & Pons, A. M. (1999). Inhibition of pathogenic *Salmonella enteritidis* growth mediated by *Escherichia coli* microcin J25 producing strains. *Can J Microbiol* **45**, 988-994.

Pugsley, A. P. (1984). The ins and outs of colicins, Part I, Production, and translocation across membranes. *Microbiol Sci* **1**, 168-175.

Qiu, X., Zhang, J., Wang, H. & Wu, G. Y. (2005). A novel engineered peptide, a narrow spectrum antibiotic is effective against vancomycin resistant *Enterococcus faecalis*. *Antimicrob Agents Chemother* **49**, 1184-1189.

Quereshi, S. T. & Medzhitov, R. (2003). Toll-like receptors and their role in experimental models of microbial infection. *Genes Immun* **4**, 87-94.

Revajova, V., Levkutova, M., Pistl, J., Herich, R., Bomba, A. & Levkut, M. (2000). The influence of colonization by *Lactobacillus* sp. and *E. coli* K88$^+$ on lymphocyte subpopulations in the peripheral blood of gnotobiotic piglets. *Acta Vet Brno* **69**, 195-199.

Riley, M. A. & Gordon, D. M. (1992). A survey of Col plasmid in natural isolates of *Escherichia coli* and an investigation into the stability of Col-plasmid lineages. *J Gen Microbiol* **138**, 1345-1352.

Riley, M. A. (1998). Molecular mechanisms of bacteriocin evolution. *Ann Rev Genet* **32**, 255-278.

Riley, M. A. & Wertz, J. E. (2002). Bacteriocin diversity: ecological and evolutionary perspective. *Biochemie* **84**, 357-364.

Sall, J., Lehman, A. & Creighton, L. (2001). *JMP Start Statistics*. Pacific Grove, CA: Duxbury Press.

Sanderson, I. R. (2004). Short chain fatty acid regulation of signaling genes expressed by the intestinal epithelium. *J Nutr* **134**, 2450S-2454S.

Schamberger, G. P. & Diez-Gonzalez, F. (2002). Selection of recently isolated colicinogenic *Escherichia coli* strains inhibitory to *Escherichia coli* O157:H7. *J Food Prot* **65**, 1381-1387.

Schamberger, G. P. & Diez-Gonzalez, F. (2004). Characterization of colicinogenic *Escherichia coli* strains inhibitory to enterohemorrhagic *Escherichia coli*. *J Food Prot* **67**, 486-492.

Schamberger, G. P., Phillips, R. L. & Jacobs, J. L. (2004). Reduction of *Escherichia coli* O157:H7 populations in cattle by addition of colicin E7-producing *E. coli* to feed. *Appl Environ Microbiol* **70**, 6053-6060.

Scharek, L., Guth, J., Reiter, K., Weyrauch, K. D., Taras, D., Schwerk, P., Schierack, P., Schmidt, M. F. G., Wieler, L. H. & Tedin, K. (2005). Influence of a probiotic *Enterococcus faecium* strain on development of the immune system of sows and piglets. *Vet Immunol immunopathol* **105**, 151-161.

Schroeder, B., Duncker, S., Barth, S., Bauerfeind, R., Gruber, A. D., Deppenmeier, S. & Breves, G. (2006). Preventive effects of the probiotic *Escherichia coli* strain Nissle 1917 on acute secretory diarrhea in a pig model of intestinal infection. *Dig Dis Sci.* **51**,724-731.

Schultz, F. & Strockbine, N. A. (2005). Genus I. *Escherichia*, Castellani and Chalmers 1919, 941T^{AL} in Bergey's Manual of Systemic Bacteriology- 2nd edition, volume II part B ed.Garitti G M, Springer, Michigan USA 607-623 pp..

Shim, S. B., Verstegen, M. W., Kim, I. H., Kwon, O. S. & Verdonk, J. M. (2005). Effects of feeding antibiotic-free creep feed supplemented with oligofructose, probiotics or synbiotics to suckling piglets increases the preweaning weight gain and composition of intestinal microbiota. *Arch Anim Nutr* **59**, 419-427.

Snoeck, V., Verfaillie, T., Verdonck, F., Goddeeris, B. M. & Cox, E. (2006). The jejunal Peyer's patches are the major inductive sites of the F4-specific immune response following intestinal immunisation of pigs with F4 (K88) fimbriae. *Vaccine* **24**, 3812-3820.

Stahl, C. H., Callaway, T. R., Linoln, L. M., Lonergan, S. M. & Genovese, K. J. (2004). Inhibitory activities of colicins against *Escherichia coli* strains responsible for postweaning diarrhoea and edema disease in swine. *Antimicrob Agents Chemother* **48**, 3119-3121.

Stein, H. H. & Kil, D. Y. (2006). Reduced use of antibiotic growth promoters in diets fed to weanling pigs: dietary tools, part 2. *Anim Biotechnol* **17**, 217-231.

Tkalcic, S., Zhao, T., Harmon, B. G., Doyle, M. P., Brown, C. A. & Zhao, P. (2003). Fecal shedding of enterohemorrhagic *Escherichia coli* in weaned calves following treatment with probiotic *Escherichia coli*. *J Food Prot* **66**, 1184-1189.

Tracka, J. & Smarda, J. (2003). Is there any function for colicinogeny in the post weaning diarrhoea of piglets? *Vet Med - Czech* **48**, 190-198.

Traunter, B. W., Hull, R. A. & Darouichi, R. O. (2005). Colicins prevent colonization of urinary catheters. *J Antimicrob Chemother* **56**, 413-415.

Tsukahara, T., Iwasaki, Y., Nakayama, K. & Ushida, K. (2003). Stimulation of butyrate production in the large intestines of weaning piglets by dietary fructooligosaccharides and its influence on the histological variables of the large intestinal mucosa. *J Nutr Sci Vitaminol* (Tokyo) **49**, 414-421.

Tuohy, K. M., Rouzaud, G. C., Bruck, W. M. & Gibson, G. R. (2005). Modulation of the human gut microflora towards improved health using prebiotics- assessment of efficacy. *Curr Pharm Des* **11**, 75-90.

Tzortzis, G., Goulas, A. K., Gee, J. M. & Gibson, G. R. (2005). A novel galactooligosaccharide mixture increases the bifidobacterial population numbers in a continuous in *vitro* fermentation system and in the proximal colonic contents of pigs in vivo. *J Nutr* **135**, 1726-1731.

Valyshev, A.V., Kirillov, V. A., Kirillov, D. A. & Bukharin, O. V. (2000). The effect of inulin on the biological properties of enterobacteria. *Zh Mikrobiol Epidemiol Immunobiol* **1**, 79-80.

Van Beers-Schreurs, H. M. G., Vellenga, L., Wensing, T. H. & Breukink, H. J. (1992). The pathogenesis of the post-weaning syndrome in weaned piglets; a review. *Vet Quarterly* **14**, 29-34.

Van den Broeck, W., Cox, E. & Goddeeris, B. M. (1999). Induction of immune response in pigs following oral administration of purified fimbriae. *Vaccine* **17**, 2020-2029.

Verdonck, F., Cox, E., Van der Stede, Y. & Goddeeris, B. M. (2004). Oral immunization of piglets with recombinant F4 fimbrial adhesin FaeG monomers induces a mucosal and systemic F4-specific immune response. *Vaccine* **22**, 4291-4299.

Verdonk, J. M., Shim, S. B., van Leeuwen, P. & Verstegen, M. W. (2005). Application of inulin-type fructans in animal feed and pet food. *Br J Nutr* **93**, S125-S138.

Verstegen, M. W. & Williams, B. A. (2002). Alternatives to the use of antibiotics as growth promoters for monogastric animals. *Anim Biotechnol* **13**, 113-127.

Vinderola, C. G., Medici, M. & Perdigon G (2004). Relationship between interaction sites in the gut, hydrophobicity, mucosal immunomodulating capacities and cell wall protein profiles in indigenous and exogenous bacteria. *J Appl Microbiol* **96**, 230-243.

Walker, W. A. & Duffy, L. C. (1998). Diet and bacterial colonization role of probiotics and prebiotics. *J Nutr Biochem* **9**, 668-675.

Wheeler, D. L., Barrett, T., Benson, D. A., Bryant, S. H., Canese K, Chetvernin, V, Church, D. M., DiCuccio, M., Edgar, R., Federhen, S., Geer, L. Y., Helmberg W, Kapustin, Y., David L. Kenton, Khovayko, O., Lipman, D. J., Madden, T. L., Maglott, D. R., Ostell, J., Pruitt, K. D., Schuler, G. D., Schriml, L. M., Sequeira, E., Sherry, S. T., Sirotkin K, Souvorov, A., Starchenko, G., Suzek, T. O., Tatusov R, Tatusova, T. A., Wagner, L. & Yaschenko, E. (2006). Database resources of the national center for biotechnology information. *Nucl Acids Res* **34**, D173-D180. http://www.ncbi.nlm.nih.gov/entrez/query.fcgi?db=Nucleotide

Yi, G.F., J. A. Carroll, G. L. Allee, A. M. Gaines, D. C. Kendall, J. L. Usry, Y. Toride, & S. Izuru. (2005). Effect of glutamine and spray-dried plasma on growth performance, small intestinal morphology, and immune responses of *Escherichia coli* K88+-challenged weaned pigs. *J Anim Sci* **83**, 634-643.

Yokoyama, H., Peralta, R. C., Diaz, R., Sendo, S., Ikemori, Y. & Kodama, Y. (1992). Passive protective effect of chicken egg yolk immunoglobulins against experimental enterotoxigenic *Escherichia coli* infection in neonatal piglets. *Infect Immun* **60**, 998-1007.

Zhao, T., Tkalcic S., Doyle, M. P., Harmon B. G., Brown, C. A. & Zhao, P. (2003). Pathogenicity of enterohemorrhagic *Escherichia coli* in neonatal calves and evaluation of fecal shedding by treatment with probiotic *Escherichia coli*. *J Food Prot* **66**, 924-930.

8.0 Appendices

8.1 Appendix-A

Fig.A-2.1 Structure of colicin gene cluster on plasmid (Riley & Wertz, 2002).

http://cbcg.lbl.gov/Genome9/Talks/riley.pdf

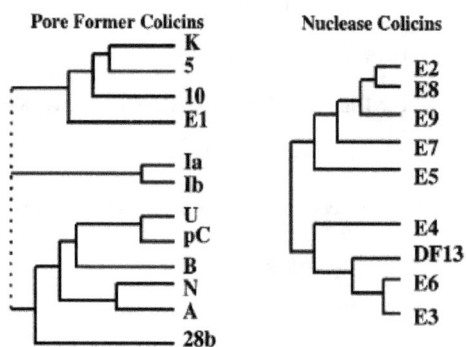

Fig.A-2.2 Phylogenetic relationship among various pore forming and nuclease colicin groups based on C-terminal 300 amino acids of colicin proteins (Riley, 1998).

http://cbcg.lbl.gov/Genome9/Talks/riley.pdf

Fig.A-2.3 The domain structure of colicins. Three domains of a colicin protein molecule are T- translocation domain, R- receptor domain and P- pore forming domain (Gökce & Lakey, 2003).

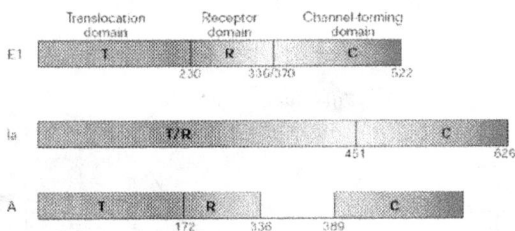

Fig.A-2.4 Domain structures and their lengths in colicins E1, Ia and A (Gouaux, 1997).

8.2 Appendix-B

Fig. B-4.1 Diagrammatic representation of modified procedure used for detection of colicinogeny in environmental isolates of *E. coli*. Procedure modified from as described by Shamberger and Diez-Gonzalez (2002).

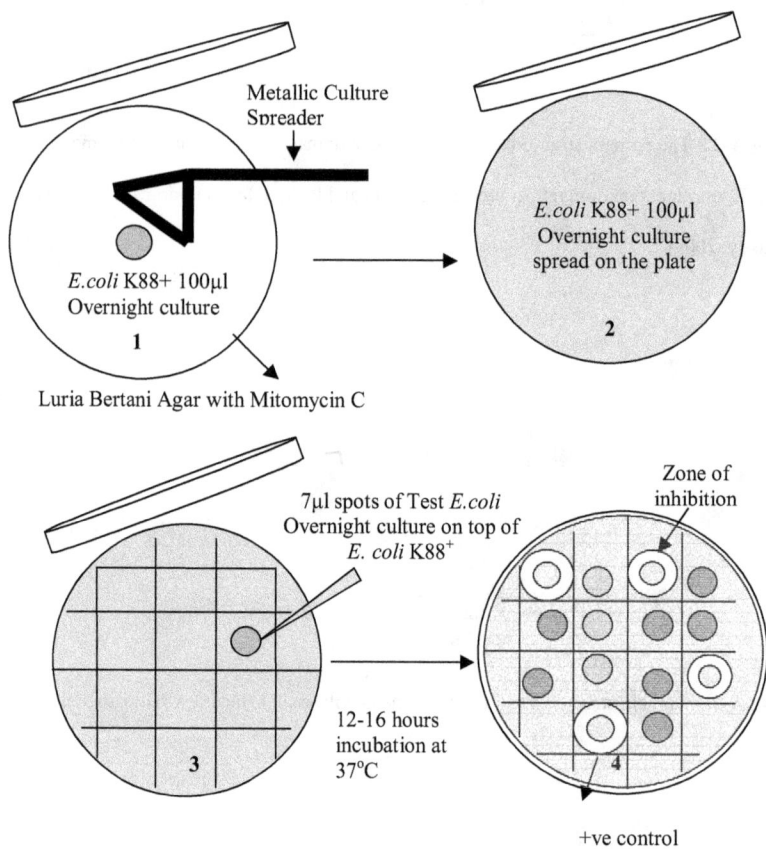

Metallic Culture Spreader

E.coli K88+ 100μl
Overnight culture

1

Luria Bertani Agar with Mitomycin C

E.coli K88+ 100μl
Overnight culture
spread on the plate

2

7μl spots of Test *E.coli*
Overnight culture on top of
E. coli K88⁺

Zone of inhibition

3

12-16 hours
incubation at
37°C

4

+ve control

Diagram by Setia A

Fig. B-4.2 Flow chart for experimental approach used in selecting potentially probiotic environmental *E. coli* isolates.

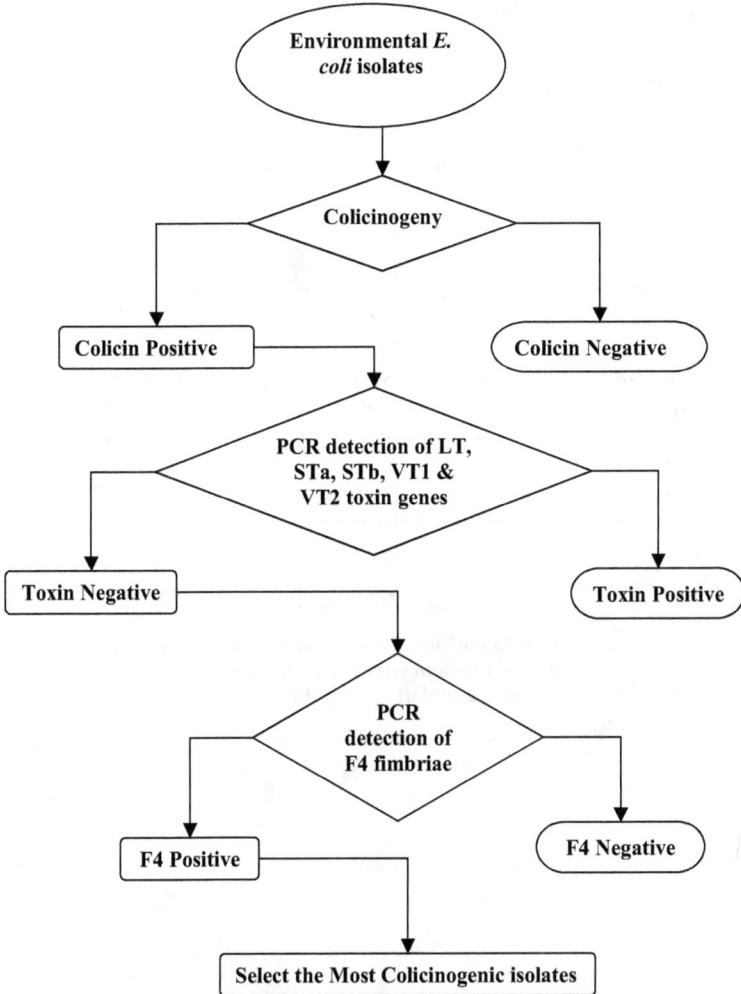

Fig.B-4.3 Characterization of 14 selected environmental *E. coli* isolates.

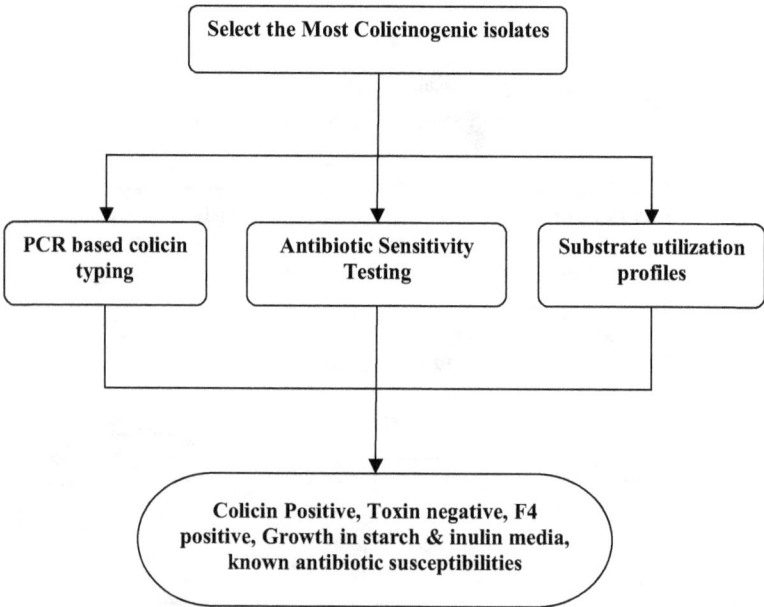

Fig.B-4.4 Plate test for detection of colicin production against *E. coli* K88[+].

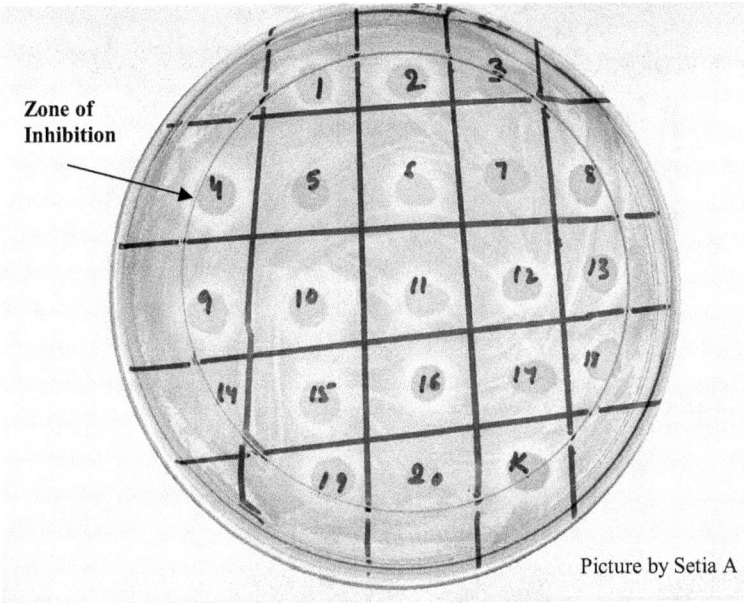

Zone of Inhibition

Picture by Setia A

Modified plate method for colicin detection is highly robust and reproducible. The method is a modification of the plate method used by Schamberger and Diez-gonzalez (2002). Colicin K producing *E. coli* K-12 construct strain BZB2116 was used as positive control for plate test. The zone around the spot was measured from the margin of the test culture spot to the end of clearance perpendicular to the start point. For irregular zones three readings around the spot were averaged. Use of this method in colicin detection is limited if confirmation is not made by methods like PCR, an actual estimation of proteinaceous substance by using Lowry's method or indirectly by elimination of activity with proteinase K treatment. PCR was used in the current study to confirm the presence of colicin genes in the fourteen selected isolates.

Fig.B-4.5 Representative pictures of agarose gel electrophoretograms.

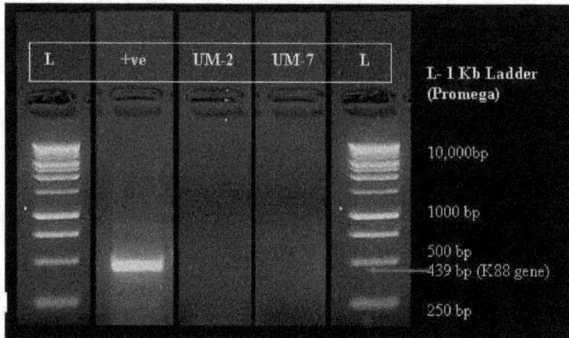

Picture by Setia A

A 2% agarose gel picture of PCR product electrophoresed for K88 (F4) gene sized 439 bp. UM-2 and UM-7 were found negative for K88 fimbrial gene as well as five toxin genes LTp, STa, STb, VT1 and VT2.

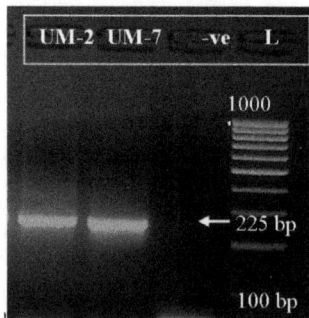

Picture by Setia A

Colicin groups were determined using PCR reactions. Isolate UM-2 and UM-7 posess genes for group A/N/S4. PCR product size for the colicin group A/N/S4 is 225 bp. UM-2 was also tested positive with colicin group B/D/D157 gene with PCR product size of 138 bp.

84

Fig.B-4.6 Standard results for API 50 CH. A. Negative reactions B. Positive reactions on API 50 CH strips. (Biomerieux) See table B-4.1 for corresponding list of substrates in each small tube on strips. Tube 0 is a blank from manufacturer.

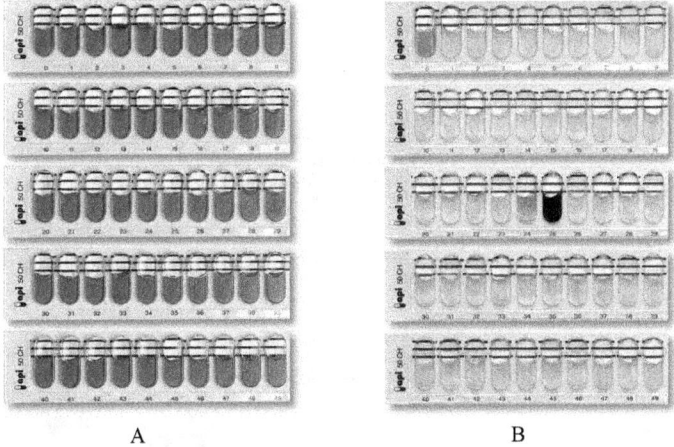

A   B

Table B-4.1 list of substrates in API 50 CH strips.

Serial no.	Active Ingredient
1.	Glycerol
2.	Erythritol
3.	D-Arabinose
4.	L- Arabinose
5.	D-Ribose
6.	D-Xylose
7.	L-Xylose
8.	D-Adonitol
9.	Methyl-βD-Xylopyranoside
10.	D-Galactose
11.	D-Glucose
12.	D-Fructose
13.	D-Manose
14.	L-Sorbose
15.	L-Rhamnose
16.	Dulcitol
17.	Inositol
18.	D-Manitol
19.	D-Sorbitol
20.	Methyl-αD-Mannopyranoside
21.	Methyl- αD-Glucopyranoside
22.	N-AcetylGlucosamine
23.	Amygdalin
24.	Arbutin
25.	Esculin + ferric citrate
26.	Salicin
27.	D-Cellobiose
28.	D-Maltose
29.	D-Lactose (Bovine origin)
30.	D-Melibiose
31.	D-Saccharose (Sucrose)
32.	D-Trehalose
33.	Inulin
34.	D-Melezitose
35.	D-Raffinose
36.	Amidon (Starch)
37.	Glycogen
38.	Xylitol
39.	Gentiobiose
40.	D-Turanose
41.	D-Lyxose
42.	D-Tagatose
43.	D-Fucose
44.	L-Fucose
45.	D-Arabitol
46.	L-Arabitol
47.	Potassium Gluconate
48.	Potassium 2-KetoGluconate
49.	Potassium 5-KetoGluconate

8.3 Appendix-C

Primer design (Benson *et al.*, 2006; Wheeler *et al.*, 2006; Hall, 1999).

Fig.C-4.1

Fimbriae K88- 439 bp

>gi|46981093|emb|AJ616256.1| *Escherichia coli* faeG gene for K88 fimbrial protein AC precursor, isolate IMM14

```
ATGAAAAAGACTCTGATTGCACTGGCAATTGCTGCACTCTGCTGCATCTGGTATG
         ATTTCAATGGTTCGGTCGATATCGGTGGTACTATCACTGCAGATGATTATCGTCAGAAATGGGA
ATGGGAAGTTGGTACAGGTCTTAATGATTGGTAATGTATTGAATGACCTGACCAATGGTGAACCAAA
CTGACCATTACTGTTACTGGTAATAAGCCAATTTGTTAGGCCGAACCAAAGAAGCATTGCTACGCCAG
TAATTGGTGGTAGATGAATTCCTCAGATTGCATTTACTGACTATGAAGGAGCTTCTGTAGAACTCAG
AAAACCTGATGGTGGAACTAATAAAGGTTAGCATATTTGTTCTGCCGATGAAAAATGCAGGGGCACT
AAAGTTGGTTCAGTGAAAGTGAATGCATCTTATGCCGGTGTGTTAGGGAG
       GGGAGCTGCTTCGCTTTTGCCGACGGTTGAGCTCTATCTTTTATGGTGGTTGCCGAGGGGTTC
       TGAACTCTCGGCTGGGAGTGCCCGAGCCGCGGGGCCACAAAGTTGTTTGGAAGTCTATCAAGAGATGATATT
```

Fig.C-4.2

Colicin E1- 397 bp

>gi|25991444|gb|AF453411.1| *Escherichia fergusonii* plasmid pColE1-EF3, colicin encoding region

```
GCTTTTATGCTGTATATAAAACCAGTGGTTATATGTACAGTATTTATTTTAACTTATTGTTTTAAAAGT
CAAAGAGGATTTATAATGAAACCGCGGTAGCGTACTATAAAGATGGTGTTCCTATGATGATAAGGGA
CAGGTAATTATTACTCTTTTGAATGGTACTCCTG        CGGAGGTGGAAAAGGAG
GCAGTAAAAGTGAAAGTTCTGCAGCTATTCATGCAACTGCTAAATGGTCTACTGCTCAATTAAAGAAAAC
ACAGGCAGAGCAGGCTACCCGGGCAAAAGCTGAGCGGAAGCACAGGCGAAAGCAAAGGCAAACAGGGAT
GCGCTGACTCAGCGCCTGAAGGATATCGTGAATGAGGCTCTTCGTCACAATGCCTCACGTACGCCTTCAG
CAACAGAGCTTGCTCATGCTAATAATGCAGCTATGCAGGCGGAAGCAGCAGAGCGTTTGCGCCTTGCGAAAGC
AGAAGAAAAGCCCGTAAAGAAGCGGAAGCAGCAGAAAAGGCTTTT
        ATTGAACGGGAGAAGGCTGAAACAGAACGCCAGTTGAAACTGGCTGAAGCTGAAGAGAAACGACTGG
CTGCATTGAGTGAAGAGCTAAAGCTGTTGAGATCGCCCAAAAAAAACTTTCTGCTGCACAATCTGAAGT
```

Fig.C-4.3

Colicin A, N, S4- 225 bp

>gi|41115|emb|Y00533.1| *Escherichia coli* ColN plasmid cna gene for colicin N

```
                                    CTTAATAAAGTGTTAGCAAATCCAAAGATGAAAGTAAACAAAT
CTGATAAGGATGCCATTGTGAATGCCTGGAAACAGGTTAATGCTAAGGACATGGCTAATAAGATTGGTAA
TCTTGGCAAGGCATTTAAGGTTGCTGATTTAGCTATAAAGGTTGAGAAAATTAGGGAAAAAAGCATTGAG
GGATACAATACTGGC                      TGAATCATGGATCATTGGTGGCGTTGTTG
CTGGAGTTGCTATTAGTTTATTCGGGGCTGTGTTGAGTTTTCTCCCAATCTCTGGACTTGCAGTTACTGC
GTTGGGGGTAATAGAATAATGACGATTAGTTACTTGTCATCTTTCATAGATGCAAATCGAGTTTCGAAT
ATAAATAACATTATATCTAGTGTTATTCGATGATATTAGTGTCAGGGGAGTATT
```

Fig.C-4.4

Colicin M- 556 bp

>gi|88770133|gb|DQ381420.1| *Escherichia coli* strain APEC O1 plasmid pAPEC-O1-ColBM, complete sequence

```
TGAAGAGGGATATGTTGAATATGCGAATGACTAATACACTGTTATAAAGGCTGCATAAAAAGGCCGGAAT
CCCGGCCCTTATATTATCGCTTACCACTTTCTTTAATGTGAATTTCACCAGGAAGCAGTATCTGGTACTC
TTTA                TTGTGAGCGACTCTCCGATGACCCACGGTGAGTGCTGGCGTTA
AAATCGTATTTATCATCATATGAACGAACAACGCCATTGTAAGTCCAGGAGCCATTGGCAGAGATAGTTA
AAGTACCTTCTGTTTCAGTGTGATATTACCAAGATATGCACCGGTAATAACATTATAATCACCAGTGGC
ATGTGTGAACTTTGTAGAAACAGGGAATGTACCTACTACACCAGATTTTATAATGTCTTTTATCTGATTA
ATTTTCATAGGGGAAATTTTAAGACCAATGTTGGCGATATTAACGCTCCTTTCAGCGGCCATTACCCATA
AATAGTGAGCAAGCGCCACAATTGGTGTAGTGACATTACCGCTCATTTGTTTCATATTCATTGAGCGATA
GTCATATTGACTTGGCGCTGGTTTTGAGAAGTGGTAAGCGTCATAAACGCTTATTCCAGGGTGAAGAACC
AGATTTCGGTCGCGGCGGTTACAGAAGTAACCAATATTTGTCGATATGAT
CCCCATGTTTTTGATGTAATCCTCAAGTTGAGTTAAAGCCTGAAGACACATGTTTGGACTCTGGAAAAA
ACTATAAAACAACCTGGACTAAAAGAGGTCCAGCACCAGGAACATGCGGTGCTGAAAGAGAAAATGCACCA
```

Fig.C-4.5

Colicin V - 400 bp

AJ223631 12606 bp DNA linear BCT 12-JUN-2006 *Escherichia coli* DNA for plasmid pCoIV-K30

```
ACGGGAGCTGTTGTAGCGAAGCCACTCGTTCAAATCAATTCTCTTGACG
                 TATAGGGGGTTGAGGGCCTCCTACCCTTCACTCTTGACTATGTTAACGATAATCATTATC
GTTAGTGTTTGTGTGTAATGGGATAGAAAGTAATGGGATAAAAAGTAATGCGATAGAAAAGAACAAAAT
TAGAGTTGTTATTTGCATTTATAATAAATGCCACCGCAATATATATTGCATTAGCTATATATGATTGTGT
TTTTAGAGGAAAGGACTTTTTATCCATGCATACATTTGCTTCTCTGCAATAATGTCTGCAATATGTTAC
TTTGTTGGTGATAATTATTATTCAATATCCGATAAGATAAAAAGGAGGATCATATGAGAACTCTGACTCTA
AATGAATTAGATTCTGTTTC              ATATTGCGATGGCTATAGGAACACTATCCG
GGCAATTTGTTGCCAGGAGGAATTGGAGCAGCTGCTGGGGGTGTGGGCTGGAGGTGCAATATATGACTATGC
ATCCACTCACAAACCTAATCCTGCAATGTCTCCATCCGGT
```

Fig.C-4.6

Colicin U and colicin Y- 243 bp

```
                         1510      1520      1530      1540      1550      1560      1570      1580      1590      1600
                         ....|....|....|....|....|....|....|....|....|....|....|....|....|....|....|....|....|....|....|....|
gi|2660583|emb|Y11823.1|  TTTGTAACGGACAGAAACCCGCCGAAGAATGGCATGCTGTGGCGAAGGACAGCTGGACGGGGCTGGTCCGGTGAATACAGGGCTGGTTAATAACGCCA
gi|9081976|gb|AF197335.1|AF197  TTTGTGAACGGACAGAAACCCGCCGAGAATGGCATGCTGTGGCGAAGGACAGCTGGACGGGGCTGGTCCAGTGAATGTAGGGCTGGTTAATAACGCCA
Clustal Consensus        *****..********** ************ **************** ******* ****** ***** ****** ***************** ********

                         1610      1620      1630      1640      1650      1660      1670      1680      1690      1700
                         ....|....|....|....|....|....|....|....|....|....|....|....|....|....|....|....|....|....|....|....|
gi|2660583|emb|Y11823.1|  TCAAAAGTGTCCGGATTATCAAAAAAGCTATGTGACGGGCGTTCTCACGCCAGAGAGGGTGATAAAGCAGGTACAAAGCAATGCGACAGGCTTT
gi|9081976|gb|AF197335.1|AF197  TCAAAAGTGTCCGGATTATCAAAAAAGCTATGTGACTGGTGTTCTTGCCTGAAGAGGTGATGAATAAAGCAGGTACAAAGCAATGCGGCAGGCTTT
Clustal Consensus        ****************************** *** **** ***** *** *** * ***** ****** ********************* ********

                         1710      1720      1730      1740      1750      1760      1770      1780      1790      1800
                         ....|....|....|....|....|....|....|....|....|....|....|....|....|....|....|....|....|....|....|....|
gi|2660583|emb|Y11823.1|  TGATCCCTTCCACTGGCAAAACAGGGAGAGGCTGTCAGACAGATTGTCGCGCCTGGACCATATCGGAGTCTCGGCATATCGAAAAAGAG
gi|9081976|gb|AF197335.1|AF197  CGATTCCCTTCCACTGGCAAAACAGGGAGAGGCTGTCAGACAGATTGTTGCAGCCTGGACTCTGGCATATCAGGATTTCCCGGTAAATCTGAAAAAAGAT
Clustal Consensus        ************************************************ *** ******* *** *** ******** *** ******* ********
```

92

Fig.C-4.7

Colicin Ia, Ib- 385 bp

```
                    1610      1620      1630      1640      1650      1660      1670      1680      1690      1700
                    ....|....|....|....|....|....|....|....|....|....|....|....|....|....|....|....|....|....|....|....|
gi|595878|gb|U15624.1|ECU15624   ATGGCGGGGCAGCCCTTGCCGTTCTTGATGCACAACAGGCCCGTCTGCTCGGGCAGCAGACACGGAATGACAGGGCCATTTCAGAGGCCCGGAATAAACT
gi|595875|gb|U15623.1|ECU15623   ATGGCGGGGCAGCCCTTGCCGTTCTTGATGCACAACAGGCCCGTCTGCTCGGGCAGCAGACACGGAATGACAGGGCCATTTCAGAGGCCCGGAATAAACT
gi|595881|gb|U15625.1|ECU15625   ATGGCGGGGCAGCCCTTGCCGTTCTTGATGCACAACAGGCCCGTCTGCTCGGGCAGCAGACACGGAATGACAGGGCCATTTCAGAGGCCCGGAATAAACT
gi|6960320|gb|M13819.2|CIAIAIM   ATGGCGGGGCAGCCCTTGCCGTTCTTGATGCACAACAGGCCCGTCTGCTCGGGCAGCAGACACGGAATGACAGGGCCATTTCAGAGGCCCGGAATAAGCT
gi|67551174|gb|AY913944.1|       ATGGCGGGGCAGCCCTTGCCGTTCTTGATGCACAACAGGCCCGTCTGCTCGGGCAGCAGACACGGAATGACAGGGCCATTTCAGAGGCCCGGAATAAGCT
gi|41141|emb|X01009.1|           ATGGCGGGGCAGCCCTTGCCGTTCTTGATGCACAACAGGCCCGTCTGCTCGGGCAGCAGACACGGAATGACAGGGCCATTTCAGAGGCCCGGAATAAACT
Clustal Consensus                * ***************************************************************************************** *****

                    1710      1720      1730      1740      1750      1760      1770      1780      1790      1800
                    ....|....|....|....|....|....|....|....|....|....|....|....|....|....|....|....|....|....|....|....|
```

```
                    1910      1920      1930      1940      1950      1960      1970      1980      1990      2000
                    ....|....|....|....|....|....|....|....|....|....|....|....|....|....|....|....|....|....|....|....|
gi|595878|gb|U15624.1|ECU15624   CAGTCATCGATAACCGTGCAAACCTGAATTATCTTCTGGCCATTCCGGTCTGGACTATAAACGCAATATTCTGAATGACCAGATCCGGTGGTGACAGA
gi|595875|gb|U15623.1|ECU15623   CAGTCATCGATAACCGTGCAAACCTGAATTATCTTCTGGCCATTCCGGTCTGGACTATAAACGCAATATTCTGAATGACCAGATCCGGTGGTGACAGA
gi|595881|gb|U15625.1|ECU15625   CAGTCATCGATAACCGTGCAAACCTGAATTATCTTCTGGCCATTCCGGTCTGGACTATAAACGCAATATTCTGAATGACCGGAATCGGTGGTGACAGA
gi|6960320|gb|M13819.2|CIAIAIM   CAGTCATCGATAACCGTGCAAACCTGAATTATCTTCTGGCCATTCCGGTCTGGACTATAAACGCAATATTCTGAATGACCAGATCCGGTGGTGACAGA
gi|67551174|gb|AY913944.1|       CAGTCATCGATAACCGTGCAAACCTGAATTATCTTCTGGCCATTCCGGTCTGGACTATAAACGCAATATTCTGAATGACCAGATCCGGTGGTGACAGA
gi|41141|emb|X01009.1|           CCGTCATCGATAACCGTGCAAACCTGAATTATCTTCTGACCCATTCCGGTCTGGACTATAAACGCAATATTCTGAATGACCGGTGGTGGTGACAGA
Clustal Consensus                * *************************************** ** ***********************************          ********

                    2010      2020      2030      2040      2050      2060      2070      2080      2090      2100
                    ....|....|....|....|....|....|....|....|....|....|....|....|....|....|....|....|....|....|....|....|
gi|595878|gb|U15624.1|ECU15624   GGATGTGGAGGTGACAAGAAAATTTATAATGCTGAAGTTGCTGAATGGGATAAGTTACGGCAAGATTGCTGATGCCAAGAATAAATCACCTCTGCT
gi|595875|gb|U15623.1|ECU15623   GGATGTGGAGGTGACAAGAAAATTTATAATGCTGAAGTTGCTGAATGGGATAAGTTACGGCAAGATTGCTGATGCCAAGAATAAATCACCTCTGCT
gi|595881|gb|U15625.1|ECU15625   GGATGTGGAGGTGACAAGAAAATTTATAATGCTGAAGTTGCTGAATGGGATAAGTTACGGCAAGATTGCTGATGCCAAGAATAAATCACCTCTGCT
gi|6960320|gb|M13819.2|CIAIAIM   GGATGTGGAGGTGACAAGAAAATTTATAATGCTGAAGTTGCTGAATGGGATAAGTTACGGCAAGATTGCTGATGCCAAGAATAAATCACCTCTGCT
gi|67551174|gb|AY913944.1|       GGATGTGGAGGTGACAAGAAAATTTATAATGCTGAAGTTGCTGAATGGGATAAGTTACGGCAACGATTGCTTGATGCCAAGAATAAATCACCTCTGCT
gi|41141|emb|X01009.1|           GGATGTGGAGGTGACAAGAAAATTTATAATGCTGAAGTTGCTGAATGGGATAAGTTACGGCAACGATTGCTTGATGCCAAGAATAAATCACCTCTGCT
Clustal Consensus                ****************************************************************          ***********************

                    2110      2120      2130      2140      2150      2160      2170      2180      2190      2200
                    ....|....|....|....|....|....|....|....|....|....|....|....|....|....|....|....|....|....|
```

Fig.C-4.8

Colicin B, D, D157- 138 bp

```
                    660       670       680       690       700
                    ....|....|....|....|....|....|....|....|....|....|....|....|....|....|....|
gi|41086|emb|X14941.1|      ATTTCGGCTGGTGATGAAACCAGGAGGCTCATCAGGTATCGCTCCATCCATCGCTCCGGGATGGGGGGATTACAGCCCACAGGGTATCGCACTTGTAC
gi|2791328|emb|Y10412.1|    ATTTCGGCTGGTGATGAAACCAGGAGGCTCATCAGGTATCGCTCCATCCATCGCTCCGGGATGGGGGGATTACAGCCCACAGGGTATCGCACTTGTAC
gi|1455661|gb|M16816.1|ECOCOLB  ATTCCGGCGCGGTGATGTGAAACCCGGAGGCTCATCAGGTCTCGCTCCATCCATCGCTCCGGGATGGGGGGATTACAGCCCACAAGGTATCGCACTTGTAC
Clustal Consensus               *** **** ****** ****** ***** ***********  *** ***** ***************** ******** ****** ********* * ***

                    710       720       730       740       750       760       770       780       790       800
                    ....|....|....|....|....|....|....|....|....|....|....|....|....|....|....|....|....|....|....|....|
gi|41086|emb|X14941.1|      AAAGTGTCTTTTCCGGGAATTATTCGCCGGATTATTCTGGATAAGGAACTTGGATAAGAACTGAAGAGGGAGACTGGTCGGGATGGTCTGTCAGTGTGCATAGCCCTTG
gi|2791328|emb|Y10412.1|    AAAGTGTCTTTTTCCGGGAATTATTCGCCGGATTATTCTGGATAAGGAACTTGGATAAGAACTGAAGAGGGAGACTGGTCGGGATGGTCTGTCAGTGTGCATAGCCCTTG
gi|1455661|gb|M16816.1|ECOCOLB  AAAGTGTCTTTTTCCTGGAATTATTCGCCGGATTATTCGCCGGATTCTGATAAAGAACTGAAGAGGGAGACTGGTCGGGATGGTCTGTCAGTGTGCATAGCCCCTG
Clustal Consensus               ***********   *****************************  **** ** ***** ** ***** ***********************  *************  **
```

gi|41086|emb|X14941.1|
gi|2791328|emb|Y10412.1|
gi|1455661|gb|M16816.1|ECOCOLB
Clustal Consensus

Fig.C-4.9

Colicin Group E (E2, E3, E3a, E4, E5, E6, E7, E8, E9) - 219bp

```
                           3710      3720      3730      3740      3750      3760      3770      3780      3790      3800
                           ....|....|....|....|....|....|....|....|....|....|....|....|....|....|....|....|....|....|....|....|
M29885.1|CE2CEALIB         GATGCTACGCATCCGGTTGAAGCGGCTGAGCGAAATTATGAACGCGCCGTGCCAGACGTGAATCAGGCAAATGAAGATGCTTGCCAGAAATCAGGAGGAGGAC
Clustal Consensus          * **  **** ** ****  ***  **  ** * ***** *** ****** *** *****  **   ***** ****    **   *  ***** *** *** ** **

                              3810      3820      3830      3840      3850      3860      3870      3880      3890      3900
                              ....|....|....|....|....|....|....|....|....|....|....|....|....|....|....|....|....|....|....|....|
AF540491.1|                AGGCTAAAGCTGTTCAGGTTTATAATTCTCGTAAAAGTGAACTGATGCAGGCGAATAAAAACTCTTGCTGATGCAAAGGCTGAAATAAAACAATTCGATCG
M62409.1|CE7CEAE7A         AGGCTAAAGCTGTTCAGGTTTATAATTCTCGTAAAAGTGAACTGATGCAGGCGAATAAAAACTCTTGCTGATGCAAAGGCTGAAATAAAACAATTCGATCG
X12591.1|                  AGGCTAAAGCTGTTCAGGTTTATAATTCGCGTAAAAGCGAACTTGATGCAGGCGAATAAAAACTCTTGCTGATGCAATAGCTGAAATAAAACAATTTAATCG
DQ916145.1|                AGGCTAAAGCTGTTCAGGTTTATAATTCGCGTAAAAGCGAACTTGATGCAGGCGAATAAAAACTCTTGCTGATGCAATAGCTGAAATAAAACAATTTGATCG
J01574.1|PIMCOLE3A         AGGCTAAAGCTGTTCAGGTTTATAATTCGCGTAAAAGCGAACTTGATGCAGGCGAGCGAATAAAAACTCTTGCTGATGCAATAGCTGAAATAAAACAATTTAATCG
X02397.1|                  AGGCTAAAGCTGTTCAGGTTTATAATTCGCGTAAAAGCGAACTTGATGCAGGCGAATAAAAACTCTTGCTGATGCAATAGCTGAAATAAAACAATTTAATCG
M31808.1|CECCOLE6A         AGGCTAAAGCTGTTCAGGTTTATAATTCGCGTAAAAGCGAACTTGATGCAGGCGAATAAAAACTCTTGCTGATGCAATAGCTGAAATAAAACAATTTAATCG
M29885.1|CE2CEALIB         AGGCTAAAGCTGTTCAGGTTTATAATTCGCGTAAAAGCGAACTTGATGCAGGCGAATAAAAACTCTTGCTGATGCAATAGCTGAAATAAAACAATTTAATCG
Clustal Consensus          *****************************  ****** **** ******  ************ ********  **** ************* ** ** **

                              3910      3920      3930      3940      3950      3960      3970      3980      3990      4000
                              ....|....|....|....|....|....|....|....|....|....|....|....|....|....|....|....|....|....|....|....|
AF540491.1|                ATTTGCCCGAGAACCAATGGCTGCTGGTCACCGAATGTGGCAAATGGCCAGGGCTTAAGGCCCAGCGGGCACAGACGGGGTAAATAATAAGGAGGCTGCA
M62409.1|CE7CEAE7A         ATTTGCCCGAGAACCAATGGCTGCTGGTCACCGAATGTGGCAAATGGCCAGGGCTTAAGGCCCAGCGGGCACAGACGGGGTAAATAATAAGGAGGCTGCA
X12591.1|                  ATTTGCCCATGACCCAATGGCTGGCGGTCACAGAATGTGGCAAATGGCCAGGGCTTAAGCTCAGGGGGCGCAGACGGATGTAAATAATAAGCAGGCTGCA
DQ916145.1|                ATTTGCCCATGACCCAATGGCTGGCGGTCACAGAATGTGGCAAATGGCCAGGGCTTAAGCTCAGGGGGCGCAGACGGATGTAAATAATAAGCAGGCTGCA
J01574.1|PIMCOLE3A         ATTTGCCCATGACCCAATGGCTGGCGGTCACAGAATGTGGCAAATGGCCGGGGCTTAAGCCCAGCGGGCGCAGACGGGGTAAATAATAAGCAGGCTGCA
X02397.1|                  ATTTGCCCATGACCCAATGGCTGGCGGTCACAGAATGTGGCAAATGGCCGGGGCTTAAGCCCGGCGGGCGCAGACGGGGTAAATAATAAGCAGGCTGCA
M31808.1|CECCOLE6A         ATTTGCCCATGACCCAATGGCTGGCGGTCACAGAATGTGGCAAATGGCCCGGGCTTAAGCCCGGCGGGCGCAGACGGATGTAAATAATAAGCAGGCTGCA
M29885.1|CE2CEALIB         ATTTGCCCATGACCCAATGGCTGGCGGTCACAGAATGTGGCAAATGGCCGGGACTTAAGCTCAGCGGGCGCAGACGGATGTAAATAATAAGCAGGCTGCA
Clustal Consensus          ********    * ****** *********  ************** *** *** * ***  *    ***  **** **   ***** ****** ******

                              4010      4020      4030      4040      4050      4060      4070      4080      4090      4100
                              ....|....|....|....|....|....|....|....|....|....|....|....|....|....|....|....|....|....|....|....|
AF540491.1|                TTTGATGCTGCTGCAAAGATAAGTCTGATGCTGATGCTGCATTGCATTGAGTCTGCGTTGGAGCCGCAAACGAAAGAAAATAAAGAAAAGGACTCTAAGG
M62409.1|CE7CEAE7A         TTTGATGCTGCTGCAAAGATAAGTCTGATGCTGATGCTGCATTGCATTGAGTCTGCGTTGGAGCCGCAAACGAAAGAAAATAAAGAAAAGGACGCTAAAG
Clustal Consensus          ********************  ******** * ** ** ** **  **  ***** ** ** * ****  ****** **  *** ****** **
```

Fig.C-4.10

Colicin 5, 10, K- 803 bp

```
                      3310      3320      3330      3340      3350      3360      3370      3380      3390      3400
                      ....|....|....|....|....|....|....|....|....|....|....|....|....|....|....|....|....|....|....|....|
gi|11124899|emb|X87834.1|        CTTAATGAAGAGAAACAGGCGGGTGGCGGAAGCAGAGAACGCTTTAGCTGAGGCGAAGCTGAACTGGCGAAGGCTGAAAGTGATGTACAGAGTAAGCAAG
gi|13532661|gb|U27452.1|ECU02745 CTTAATGAAGAGAAACAGGCGGGTGGCGGAAGCAGAGAACGCTTTAGCTGAGGCGAAGCTGAACTGGCGAAGGCTGAAAGTGTACAGAGTAAGCAAG
gi|61229209|gb|AY929248.1|       CTTAATGAAGAGAAACAGGCGGGTGGCGGAAGCAGAGAACGCTTAGCTGAGGCGAAGCTGAACTGGCGAAGGCTGAAAGTGATGTACAGAGTAAGCAAG
gi|12128921|emb|X87835.1|        CTTGATGAAGAGCATCGGGCTGTGAAGTGGCCAGGAGAAGCTGGCTGAGGCTAAAGCTGAACTGGCGAAGGCCGAAAGCGATGTACAGAGTAAGCAAG
gi|8078751|emb|X82682.1|         CTTGATGAAGAGCATCGGGCTGTGAAGTGGCCAGGAGAAGAAGCTGGCTGAGGCTAAAGCTGAACTGGCGAAGGCCGAAAGCGATGTACAGAGTAAGCAAG
Clustal Consensus                ***  ******** *  **** ****            *  **********  *     ****** ********


                      3410      3420      3430      3440      3450      3460      3470      3480      3490      3500
                      ....|....|....|....|....|....|....|....|....|....|....|....|....|....|....|....|....|....|....|....|
                      4110      4120      4130      4140      4150      4160      4170      4180      4190      4200
                      ....|....|....|....|....|....|....|....|....|....|....|....|....|....|....|....|....|....|....|....|
gi|11124899|emb|X87834.1|        CTGGGGGATTGCAATTGTCACAGGTATTGTTTCTTCTTACATAGGGGATGAGAGTTGAGCAAACTTAATGAGTTGTTAGGGATTTAATTTCTCTTTAGA
gi|13532661|gb|U27452.1|ECU02745 CTGGGGGATTGCAATTGTCACAGGTATTGTTTCTTCTTACATAGGGGATGATGAGTTGAGCAAACTTAATGAGTTGTTAGGGATTTAATTTCTCTTAGA
gi|61229209|gb|AY929248.1|       CTGGGGGATTGCAATTGTCACAGGTATTGTTTCTTCTTACATAGGGGATGATGAGTTGAGCAAACTTAATGAGTTGTTAGGGATTTAATTTCTCTTTAGA
gi|12128921|emb|X87835.1|        CTGGGGGATTGCAATTGTCACAGGTATTGTTTCTTCTTACATAGGGGATGATGAGTTGTAGGGATTTAAGTGTTAGGGATTTAATTTCTCTTTAGA
gi|8078751|emb|X82682.1|         CTGGGGGAATTGCAATTATCACAGGTATTGTTTCTTCTTACATAGGGGATGATGAGTTGAACAAGCTTAATGAATTACTAGGTATTTAATTTCTCTTTAGA
Clustal Consensus                ******** *** ** ******************************   *** ***  **   *  ***   ***   ** ****************** **
```

www.ingramcontent.com/pod-product-compliance
Lightning Source LLC
Chambersburg PA
CBHW061609220326
41598CB00024BC/3503